COSMIC REALITY

UNDERSTANDING SPACE, TIME, AND EINSTEIN'S UNIVERSE

VRUSHANK CHARAN AGRAWAL

INDIA • SINGAPORE • MALAYSIA

Notion Press

Old No. 38, New No. 6
McNichols Road, Chetpet
Chennai - 600 031

First Published by Notion Press 2019
Copyright © Vrushank Agrawal 2019
All Rights Reserved.

ISBN 978-1-64733-565-6

This book has been published with all efforts taken to make the material error-free after the consent of the author. However, the author and the publisher do not assume and hereby disclaim any liability to any party for any loss, damage, or disruption caused by errors or omissions, whether such errors or omissions result from negligence, accident, or any other cause.

While every effort has been made to avoid any mistake or omission, this publication is being sold on the condition and understanding that neither the author nor the publishers or printers would be liable in any manner to any person by reason of any mistake or omission in this publication or for any action taken or omitted to be taken or advice rendered or accepted on the basis of this work. For any defect in printing or binding the publishers will be liable only to replace the defective copy by another copy of this work then available.

In the loving memory of my uncle

Contents

Preface *7*

Acknowledgments *13*

Workings of Space and Time

Chapter 01 The Modern Scientific Age 17

Chapter 02 An Intro to the Special Theory 37

Chapter 03 The Weirdness Begins 47

Chapter 04 Playing with Length & Time 63

Chapter 05 Paradoxes to Ponder 75

Chapter 06 Special Theory in Real Life 91

Chapter 07 What Is so General about Relativity 107

The Cosmos in a Platter

Chapter 08 From the Beginning 127

Chapter 09 Stellar Evolution – Dwarfs to Giants 143

Chapter 10   Holes & Waves in the Dark Cosmos...... 157

Chapter 11   Time, Infinity and Beyond...................... 193

Glossary..*223*

References..*231*

Preface

Physics is a mysteriously fascinating subject. It is like an elegant conception of the human mind, which has been created out of the unknown to describe the unknown. The roots of modern physics can be traced back to more than three millenniums ago or even before when numbers and basic mathematical ideas were created. The sole purpose behind those innovations was to formalize everyday actions such as trade and decree to make them more systematic, uncomplicated, and comprehensive than they initially were. Further down the road, the Greeks with the Romans invented philosophy, one of the most ancient forms of academia that involves the general study of fundamental topics such as knowledge, life, mind, existence, and consciousness. Both of these prime academic subjects were born out of the curiosity of the human mind to explain the everyday phenomenon and answer profound questions about life like who we are, and what existence means. Since their origin, both of these profound subjects, which are like two parallel roads, have evolved over the centuries while finetuning our species' understanding of nature. On a dark rainy day somewhere in between all of these centuries, in an unprecedented event, somehow, these two parallel roads dissected, turned, and merged to produce a third separate road, which we today call physics.

The arena of physics is the study of the nature of the cosmos. It involves notions from philosophical understandings and employs mathematical concepts to explain fundamental and intriguing concepts like space and time. It is no surprise that these properties of the universe have captivated and motivated generations of mathematicians, philosophers, and scientists to learn more about them. For eventually, every action, event, or phenomenon which takes place in the universe, does so on the fabric of space and the flow of time. This feature makes these concepts – space and time – so profound and unequivocally significant that their understanding remains integral to the upliftment of our society. The technological breakthroughs and theoretical improvements that we achieve today are, in some way or the other, fundamentally linked with these two essential concepts.

The ironical fact, though, is that even after hundreds of years of incremental work in this field and regular upheaval of longstanding theories to make way for better, more accurate theories, our generation is principally at the same junction where Aristotle's generation was during their time. Their definition of the universe worked in almost all aspects but some; similarly, our description of reality works in nearly every aspect but some, where phenomena remain unexplained even after using the best tools we have at hand. Essentially, the only difference from Aristotle and now is in the concepts and models which have been employed to explain that equivalent reality. They used gods; we did not. They used heavens; we did not. Primarily, the scale of our understanding may be much larger than Aristotle's generation, our discoveries may be more widespread than theirs, but still, the fundamental questions which led their generation to innovate and will lead our generation to innovate in the future, are the same. They were inquisitive and eager to know what reality, space, and time are? After 2000 years of scientific work, we too are eager and inquisitive to learn, but still, we ask the same questions as Aristotle's generation did. That is ridiculous. The sad truth is that our theories, even after all these centuries, have not been able to produce a definitive answer to those fundamental questions. All we have are speculations, which, although different today, even Aristotle had. Predominantly, the questions we ask today have been reinterpreted in various disguises from previous generations and will be reinterpreted time and again in the future; until one day, when a generation, probably the *billennials*, come across a conclusive answer, which could potentially explain everything there is to explain without contradictions, untestable predictions, or further possibilities for improvement.

Nevertheless, satisfactory if it may be, the truth is that to look for this answer to the secret of life is like looking for one specific atom in a universe of particles, where observations and theories are all that we have to speculate upon; hence, we must take them seriously. For a good reason too. Today's theories are not aided by unguided imagination and baseless skepticism; instead, they are guided by mathematical frameworks and philosophical roots that make them more far-reaching and emulating than ever before. For example, Aristotle believed that the sun revolves around the Earth, but we know better, it is the Earth that revolves around the Sun. Newton believed in the existence of absolute space, but we still know better, there is no absolute space. Primarily, the questions which fuel our generation's passion towards achieving scientific breakthroughs may not have significantly changed, but the paradigm shifts that we have encountered as well as the refinement in ideas that we have produced, have been laudably positive. Furthermore, these achievements are enough for us to keep the good work going while simultaneously keeping us optimistic about future innovations and discoveries, which may lead us to more profound world views of the universe.

In this book, we will be looking through the most popular and widely accepted idea of reality present in the scientific world today. This idea revolves around Einstein's theory of relativity, which regrettably remains mostly obscure to individuals in the conventional world. Einstein's thoughts on space, time, and reality, in essence, are only improvements over Newton's ideas, but yet, they have been physics-changing in almost every aspect imaginable. There are further tacks at explaining the reality that go beyond Einstein, but these concepts are still theories in the making as their existence is limited to hypotheses on paper but not experimental proofs in laboratories. Einstein's revolutionary disruption of Newton's classical ideas will hence, largely frame our passage in the pages to follow while making us cognizant about the most widely agreed upon modern-day understanding of the nature of the universe.

The book is intended mainly for high school students who have a deep craving to learn about the mechanisms of the universe, a topic that has fascinated eager individuals across millenniums. The first part of the book introduces theoretical concepts of Einstein's theory of relativity through illustrations, analyses, and almost mathematical-less explanations, whereas the second part of the book introduces the macroscale implications of the concepts introduced in the first part. These concepts and implications which have been presented are, however, only a small part of the entire theory because the rest of the unintroduced ideas require advanced mathematics and physics for their complete comprehension.

Apart from high school students, the book will also be merchandise of interest to avid readers, science enthusiasts, and other individuals who, although yearn to know about the nature of the cosmos, have no significant academic background in the field of physics. Since, the book revolves around core concepts that pave our way towards an accurate understanding of the universe while also elucidating on the reasons why Newton's ideas are outdated and hence, a significant update is required in our discernment of reality. Moreover, the potentially new concepts discussed in further chapters have been interlinked with popular everyday perceptions that the general observer, in all fairness, is expected to be familiar with in the form of metaphors and analogies. Some paragraphs, which provide additional information regarding the subject matter, have been outlined in the book to forewarn the reader if he/she may decide to skip that part, though, the reader is advised against doing that.

Many ideas presented in the book are controversial, but it has been attempted to offer both sides of the story in such cases with complete neutrality while leaving the power of decision making with the reader based on the facts presented. A short glossary, after the main content, provides definitions of the critical scientific terms used through the course of the book. Also available, after the main content, is an afterword, which provides the curious reader an opportunity to learn about the *'other tacks at explaining reality'* mentioned earlier. Furthermore, every chapter begins with words of wisdom from celebrity scientists, which, in most cases, are related to the subsequent matter present in the section. The diligent reader is encouraged to appreciate these quotations by stressing upon their meanings, which may be more profound than the reader may realize at first sight considering the context of the entire subject in general.

Concluding our brief visit through this long and informative introduction, and conforming to the words of Stephen Hawking, *"My goal is simple. It is the complete understanding of the universe, why it is as it is, and why it exists at all,"* we henceforward embark on our unique voyage: *understanding the cosmos*.

Acknowledgments

I want to thank my parents for their untiring support, contribution, and motivation throughout my journey through physics, without which I would not have been able to publish this book. I also want to thank all of my physics teachers and friends, familiar with the subject matter, for their treasured feedback and suggestions, which largely helped me shape the content of the chapters and streamline my thoughts to precision. A special thanks to my younger sister for proofreading significant parts of the raw manuscript that helped me improve the overall grammatical framework. The fact that her age was among that of the target audience facilitated me immensely in rationalizing the concepts in the book to make sure that they were delivered to the reader with the right emphasis and simplicity while also being compliant with the core principles of the ideas presented.

Workings of Space and Time

> "There are only two ways to live your life. One is as though nothing is a miracle. The other is as though everything is a miracle."
>
> – Albert Einstein

CHAPTER 01

The Modern Scientific Age

"Science is a way of life. Science is a perspective. Science is the process that takes us from confusion to understanding in a manner that's precise, predictive, and reliable – a transformation, for those lucky enough to experience it, that is empowering and emotional."

– Brian Greene

"The most beautiful thing we can experience is the mysterious. It is the source of all true art and science."

– Albert Einstein

"Whence come I and whither go I? That is the great unfathomable question, the same for every one of us. Science has no answer to it."

– Max Planck

"Modern science says: 'The sun is the past, the earth is the present, the moon is the future.' From an incandescent mass we have originated, and into a frozen mass we shall turn. Merciless is the law of nature, and rapidly and irresistibly, we are drawn to our doom."

– Nikola Tesla

Over the past two and half millennium's scientists, philosophers, and mathematicians have valiantly tried to consolidate our understanding of nature through geometry, algebra, physical sciences, and philosophy. The earliest scientific studies focused on the characters of objects and phenomena one might see or experience in everyday life. Shouting *'Eureka'* while running on the streets of Syracuse with only a towel around his waist, Archimedes discovered the *'Archimedes Principle,'* which set the basis of fluid mechanics widely used by engineers even today. Plato played a pivotal role in establishing western philosophy, and his work has served modern philosophy even until the early 20th century. Aristotle, Plato's student, contributed to a myriad of fields, including physics, biology, zoology, psychology, among many others, and is famously remembered as the *"Father of Western Philosophy."*

The majority of scientific work done by men during the past two millenniums was lost due to warfare or inadequacy in its preservation. The work which was preserved was given little attention as it largely remained obscure to the general public. It was after the invention of the printing press in the mid-1400s that the wide dissemination of incremental work in the field of science, math, geometry, astronomy, and philosophy became commonplace. The statement is debatable, but it is also widely agreed upon that modern scientific age began by the time of Kepler and Galileo before the rise of Newton, who went on to initiate a revolution. The goal of their investigations and experiments was to fabricate theories that could satisfactorily rationalize the regularity of the events which took place commonly in the physical reality in which they lived and experienced.

As legend has it, Galileo studied weights falling from the leaning tower of Pisa while Newton investigated dropping apples and the revolving moon. Several other sung and unsung heroes played essential and indispensable parts in the progress of Physics and Mathematics in the 1600s, but Sir Isaac Newton, the first-ever scientist to be knighted, undoubtedly stole the show. Newton invented the idea of gravity, which was an integral insight in his publication *Philosophiae Naturalis Principia Mathematica*, considered to be one of the most important single works in the history of modern science. His work laid down the backbone of physics, mainly classical mechanics, by introducing the idea of motion and gravity. The universal law of gravitation explained how the planets revolve in specific orbits around the sun, why apples fall, and also established the omnipresent nature of gravitational force exerted by each object in the cosmos.

Newton's work marked the epoch of a great revolution as it accentuated scientific work, which until then, remained in the darkness of hypotheses and conjecture. He remained an exemplary throughout the 1600s and the 1700s because of his invaluable contributions to the fields of science and Math. Although famously regarded for the discovery of gravity, laws of motion, and calculus, Newton had innumerable other contributions in optics and the establishment of mechanics in particular.

Newton's equations and concepts were developed into an elaborate mathematical structure in the decades to follow, and classical physics, as it is referred to, gradually grew into a sophisticated mature discipline during the 1700s. His equations, which provided a wealth of physical phenomenon through a single theoretical framework, even today, after three hundred years, are widely used. From explaining the worldview to young minds through introductory chalk-boards to being scrawled up on NASA flight plans mapping spacecraft trajectories, Newton's work is integral to the evolution of modern science and continues to form the mainstay of forefront research by being embedded in the most fundamental concepts used for more advanced understanding of workings of the cosmos. In simple words, classical or Newtonian physics provides a rigorous and firm ground for human intuition and technology development.

Newton's theories, though, majorly revolved around gravity, motion, and dynamics. It was only in the 1860s that James Clerk Maxwell, a brilliant Scottish scientist, extended the framework of classical physics to take account of electrical and magnetic forces. Maxwell integrated previous works by scientists like Faraday, Ohm, and others into his publication *"A dynamical theory of the Electromagnetic field"* in 1865. Though Maxwell required additional sophisticated mathematics to explain the electromagnetic phenomena, he observed that his equations, known as Maxwell's equations, were each bit successful in explaining the electromagnetic phenomenon as were Newton's equations to describe motion. Maxwell's work is also widely regarded as the second grand unification in physics after Isaac Newton's realization, and he is considered to be the third most influential scientist ever only behind Newton and Einstein.

Maxwell's major work on electromagnetism is a set of 4 law's which are – Gauss's Magnetic Field Law, Gauss's Electrical Field Law, Ampere-Maxwell Law, and the Faraday

Law. Respectively, the first law establishes a relationship between magnetic poles and magnetic field, the second law emphasizes on the relation between charge and electric field, the third law describes magnetic fields which are produced by electrical fields, and the fourth law (Faraday's Law) defines the electrical field which arises as a consequence of magnetic field. All these laws are inter-related and form the basis of Maxwell's work, and consequently, the equations that arise, which obey the fundamental relationships between the laws, govern the electromagnetic world and form the backbone of today's telecommunication technology.

After the works of Newton and Maxwell, a growing sense prevailed around the world during the 1890s which suggested that theoretical physics would soon become a finished subject as *"most of the grand underlying principles have been firmly established,"* remarked Albert Michelson, a renowned experimental physicist in 1894. There were minor flaws and contradictions inside the grand framework built in classical physics, but it was firmly believed that the exponentially growing contributions from scientists around the globe would soon address these *"minor details"* as Lord Kelvin regarded them.

The details mentioned above were duly addressed in the next decade, and thankfully so. However, they proved to be anything but minor as the subject which was believed to be on the brink of completion was entirely rewritten. Previous notions of the universe and its workings were thrown out of the window as a momentous scientific revolution struck the world in the early 1900s led initially by Albert Einstein's insights.

While contemplating electricity, magnetism, light's motion, among other concepts, Albert Einstein, regarded as the direct successor to Newton, realized that Newton's most fundamental belief about the nature of the universe was itself flawed. Between the years 1905 and 1915, when Einstein produced his theory of relativity, he jolted the world by establishing that the Newtonian perspective, after all, was not correct. He identified the significant shortcomings in the understanding of the workings of the universe, and addressed them by publishing his theory of relativity. Furthermore, his paper instigated the relativity revolution, similar to the one Newton's theory of gravity did a couple of centuries ago. Einstein's work, one of humanity's most precious achievements, transformed the way the universe was classically understood as he rewrote physics' laws, which have since helped our civilization to evolve in leaps and bounds.

Nevertheless, what is this theory of relativity which crossed away from the pleasing worldview of classical Newtonian mechanics that was firmly aligned with extensive general experience?

The entire theory of relativity resembles a building consisting of two separate levels, the special and the General Theory. The Special Theory, which is integral to the General Theory, applies to all physical phenomenon except gravitation and gravitational fields. The general theory of relativity provides the law of gravity and how it is intrinsic to the very fundamentals and structure of the universe. However, to grasp the nature of both theories, it is crucial to becoming acquainted with the base of their two primary principles.

The first principle of the theory, in simple words, is: the laws of physics do not change in any frame of reference in uniform translatory motion (motion without acceleration and rotation). This principle is integral to the theory because it evades the possibility of the presence of presumed features of the universe, such as the aether (discussed later in the chapter).

The second principle of the theory is the '*principle of constant velocity of light in vacuo.*' This principle springs from the success of Maxwell's works in electrodynamics and asserts that the velocity of light in a vacuum is irrespective of the velocity of any other object and remains the same throughout (approx. 300,000 km/s). To understand this, imagine that Geeta is traveling in a spaceship at a speed of 150,000 km/s while a light beam is flashed parallel to her spacecraft. According to a classical perception and by applying Newton's laws, one concludes that Geeta would record the speed of light beam as 150,000 km/s, but the principle of light constancy asserts that she will still record the speed of light as 300,000 km/s.

I assure you that both of these principles, which seem unintuitive and logically irreconcilable at first, are powerfully supported by age-old experience and numerous experiments which we will explore in detail in further chapters. Although to successfully apply these principles in real life, one had to forgo the past laws of kinematics in Newtonian mechanics, and ultimately, the doctrine of laws relating to space and time had to be overhauled. Einstein and others did arrive at a new law of motion for rapidly moving points, confirmed in the case of electrically charged particles. Subsequently, physics had to re-adapt because Euclidean geometry could no longer be used and was replaced by non-Euclidean geometry. Moreover, any mechanical problem which was previously studied in a 3-dimensional environment had to be now evaluated in a 4-dimensional space with the fourth dimension being time itself, as time was no longer an independent quantity (chapter 3). This revolutionary change in physics has been confirmed experimentally time and again, and even today, scientists continue to confirm various untested and previously unproven predictions laid down as solutions to Einstein's equations.

One may debate, especially a high school student, that why then do we need to study Newtonian mechanics when it quite clearly does not seem to portray the world correctly, and thus, not fulfilling the fundamental essence of physics which is the scientific study of how the universe works and its explanation. Well, for one reason, Newtonian mechanics gives us reasonably accurate answers to problems we deal with in our everyday life. It is only on large cosmic scales when we deal with light speeds and consider gravitational forces between large astronomical bodies that Newtonian mechanics can no further be used to explain the phenomenon, and we need Einstein to help us out. Secondly, the theory of relativity is a little confusing, given its counter-intuitive approach towards categorizing the universe. The concepts of the theory of relativity are completely abstract as they contradict our everyday life experiences. The theory is so otherworldly that even after 100 years, physicists find it next to impossible to even visualize it.

In scientific writing, Einstein introduces his theory by saying, *"The theory of relativity' is a 'principle theory,' this means that the theory employs analytical methods rather than synthetic methods for its establishment and working. In simpler words, the basis of the theory is not hypothetically constructed but empirically discovered like the general characteristics of natural processes or the elementary principles, which give rise to mathematical formulations that are then required to be fulfilled by separate theories. The advantage of such a theory is its logical perfection and the security of its undisputable foundations."*

This introduction is relatively complex, and I by no means expect you to understand each word of it. Nevertheless, a key takeaway for us is that Einstein's theory, in a way, guards itself against anomalies and probable discrepancies because its base arises from the underlying phenomenon and fundamentals of nature that have been widely accepted.

To a critique, though, the debate looms large. He asks – whose work is more significant, Einstein's or Newton's?

Considering the question and replying to the several critiques who question the credibility of Newton's contributions after comparing them with his, Einstein says, *"Let no one suppose, however, that the mighty work of Newton can be superseded by this or any other theory. His great and lucid ideas will retain their unique significance for all time as the foundation of our whole modern conceptual structure in the sphere of natural philosophy"*. This quote or more of a response from Einstein should ably settle down the debate and put the critiques at rest.

To the Miracle Year

Newton is famous for his contributions to mechanics, but less known is the fact that he had a theory on light as well, corpuscular theory of light as he called it. In this theory, Newton presents light as a stream of particles or rather corpuscles traveling in straight lines. However, because it was presented rather sophisticatedly, this theory was not as successful as Newton's other works and is, therefore, less talked off. Fast-forwarding to the 1800s, a young scientist named Thomas Young surprised the scientific community by proposing a theory of light backed by experiments on interference in which light behaved not like a particle but as a wave. This theory was hard to believe given that it opposed Sir Isaac Newton's insights, considered as the greatest scientist to have ever lived.

Fast-forwarding to the latter half of the 1800s, scientists had by then begun noticing the effects of electromagnetism, where they found an interesting inter-relation between electric and magnetic forces where both of them creditably complemented each other. The phenomenon discovered was that an electric current could influence a magnet, which in turn could be used to generate electrical current. This phenomenon is famously known as electromagnetic induction, and a well-known experiment straightforwardly confirms it.

Take a loop of wire, pass an electric current through it, and place a compass needle near it. The observer will observe that the compass needle deflects until the electric current in the loop is switched off. In another experiment, take a loop of wire (with a small bulb attached to it) and a bar magnet, by bringing the magnet closer to the loop or even by pulling it away, one can observe that the light bulb actually glows, and as soon as the bar magnet is stopped moving, the light bulb stops glowing. So, somehow, the movement of the magnet, as it moves toward the loop, and the changing magnetic field generates an electric current. Therefore, this experiment establishes an inter-relation between electricity and magnetism.

Electromagnetic Wave

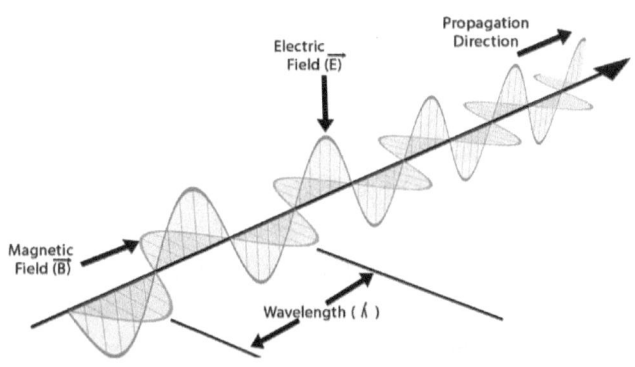

Figure 1.1 The vertical wave represents the electric wave and the horizontal wave represents the magnetic wave. Both the waves combined are called an electromagnetic wave. The line of propagation of the combined wave and the line of propagations of both separate waves are always mutually perpendicular.

This phenomenon of electromagnetic induction has since been potently exploited through the invention of the electric motor and the electric generator, but a question springs out – how does one explain this interconnection? In the 1860s and 70s, as discussed above, James Clark Maxwell put together decades of published work on electric current and magnetism, integrated them, and synthesized separate equations that accurately explained each property of the weird interconnection of electricity and magnetism. One astonishing but critical take away from his work was the prediction of the existence of electromagnetic waves. This wave consists of two separate parts, the electric and the magnetic, which travel mutually perpendicular to the plane of propagation of the wave (shown in figure 1.1). Another surprising discovery was that these electromagnetic waves seemed to travel very close to the known speed of light. Furthermore, it could be deduced by natural implication, that light may be an electromagnetic wave itself. This result was in tandem with the results that Thomas Young had produced decades ago, where he showed that light behaved not as a particle but a wave. The scientific community eventually accepted Young's theory and established that light does exhibits wave properties. This finding though poses a vital question. Water waves travel through a medium which is water, sound waves travel through a medium which is air, but through which medium do electromagnetic light waves travel?

It was, hence, proposed that an invisible substance filled the entire universe and was solely responsible for acting as a medium for transporting electromagnetic light waves. This substance or a medium, as it was thought to be, was given the name aether which in Greek personified the 'god of upper air and light of the brightly glowing heaven.' This hypothesis of the presence of aether was extremely unexpected but convincing because, on paper, it fluently explained the supposed nature of electromagnetic waves. This realization prompted the British scientists to come up with complicated mechanical models filled with sophisticated mathematical analysis to explain the aether. The French derided these models, believing that the British were off in a very wrong direction and came up with their factory models to explain the aether. But even after a lot of investment of time, energy, and money, none of the models were compelling and would always end up contradicting other established fundamental principles of Physics. Eventually, the nature of the aether remained obscure to scientists for a very long time.

It was only in 1887 that a study was finally carried out on the aether through the Michelson–Morley experiment. Although the real analysis is quite long and tedious, the basic idea behind it turns out to be relatively simple. We know that the light waves are going to travel through this aether. Now, as the Earth moves around the sun and even when it rotates on its axis, it moves through the aether. Therefore, at the surface of the Earth, we should have an aether breeze in our face, similar to a breeze we feel when we ride a bike. Now, if you send two light beams, one against that aether breeze versus one in the direction of the breeze, you should get a difference in speeds due to retardation and acceleration provided to those respective light beams by the aether, and it should be detectable. Nevertheless, after conducting this experiment, what did the experimentalists find? They found absolutely nothing! They found there was no aether breeze effect or aether wind effect, which was indeed an astonishing result. Michelson and Morley thought they had done something wrong and repeated the experiment numerous times but always ended up with the same 'null' results. Many other scientists who heard of these 'null' results said: *"This is terrible because all the aether theories seem to be wrong at this point if the experiment is true."* Scientists like Lorentz, Poincare, Fitzgerald, who proposed different theories and ways to conduct these experiments, went on working in the context of how does one construct the model of the aether when it clearly showed no signs of its presence.

At this point, Einstein comes into the picture. Let us look at the past events of his life before learning about his contributions in the next chapter.

Einstein's early life was full of hardship and struggle. He could not speak until he was four, which is rare, and most of his teachers considered him to be retarded or mentally handicapped. However, Einstein did not care about what others thought of him. In the morning, he would argue with his history teacher about the worth of remembering useless dates when he could always look them up in the textbook, while in the evening he would be fascinated by a compass trying to figure out what invisible forces and laws of nature made that tiny little needle move around. This curiosity, as Einstein says, is what fueled his imagination and motivated him to discover the laws of nature and understand the universe. Throughout his childhood, when most people around him would question his capabilities and doubt his future, Einstein would sit flat on a garden and study Newton's laws and geometry. Several myths state that Einstein failed math in school and was one of the reasons he left Germany, but none are true. In fact, Einstein learned Calculus himself, and by the age of 16, he had already mastered it. Even Einstein's high school math teacher accepted that Einstein was much better at math than him and also capable enough to teach.

By the age of sixteen, Einstein felt that he had studied physics and mathematics to the best of his capabilities with the resources he had at hand and should now aim to enter college to pursue higher education. He knew that school education was not for him. He so vehemently despised the German schooling system because of its extreme rigidity, hierarchy, and militaristic Prussian slant that he eventually gave up his German citizenship at the age of 17. At the same age, Einstein wished to shift to Switzerland and attend the Federal Polytechnic, but due to his lack of in-depth knowledge in subjects other than Mathematics, he failed to make the cut for admissions. On the advice of the principle, Einstein first obtained a diploma at the Cantonal School in Aarau before joining FIT in 1896, where he realized that Physics was better suited to him for a career than did Mathematics.

Einstein always wanted to be a Swiss citizen, and after attending the Swiss Federal Polytechnic for his undergraduate degree, he finally did receive his much-awaited citizenship. At the university, Einstein was being trained to become a high school physics teacher, and he did not oppose it because he was not sure or ambitious about his career. He graduated in 1900 with not so excellent grades but with a clear intention to find himself a job. Einstien wanted to become an assistant professor of physics at a university to explore and perform research, but most of the applications he submitted across Europe got rejected while others never got a response back. Besides, he also wanted to marry a girl, Mileva Maric, whom he had met during his time at the FIT in Switzerland. Moreover, his father had already gone bankrupt twice, which made him desperate as he felt the urgency of the need to earn money. Eventually, all of these events turned into anxieties and took a toll on Einstein, who had already gone out of touch from his career with the feeling of loneliness getting more and more entrenched in him by each passing day. Fortunately for him, in 1902, Einstein finally did get a job at the Swiss Patent Office in 1902 through an old friend's connection from the university. Although it was not a job which Einstien was looking forward to, he still accepted it given his desperate situation.

At the patent office, Einstein made a friend Michel Besso, an electrical engineer and his former classmate at the Federal Polytechnic. Besso would often accompany Einstein for long walks during lunch hour and in the evening to discuss the newest physics problems and developments in the scientific word. They would ponder and talk about the conundrums involved in the whole electromagnetic and aethereal world view. Einstein even called Besso the best *sounding board* in all of Europe and credited him for providing many valuable suggestions.

During his time at the patent office, Einstein used to conduct numerous thought experiments where he would consider highly implausible situations to understand how space and time work. He would scribble his notes, thoughts, and equations on a notepad he possessed, and whenever a senior would visit him, he would fling the pad into the top drawer of his table and turn back to his usual stamping on a pile of patent demanding documents kept in front of him. Had someone checked Einstein's desk drawer during those early years, Einstein might have never been able to complete his work apart from also being kicked out of the patent office.

In 1905, Einstein published four research papers on separate topics, with each article being revolutionary in distinct ways. These publications were not Einstein's first, but they specifically provided him recognition amongst the scientific community at the international level and that too at such an early age of 26. Such has been the contribution of those four publications to physics that if a scientist could publish even one paper of the level of what Einstein published in that single year, the achievement would have still been considered career-defining. The year 1905 itself came to be known as the *'Miracle Year'* (*annus mirabilis* in German) because that was the turning point for not only Einstein but an entire generation.

Einstein's first paper of the year was on the Photoelectric effect, which came out in March 1905. Einstein hypothesized that light behaves not like a wave but like a particle that could be used to explain the Photoelectric effect widely observed but unexplained for over five decades. Einstein's mathematical equations in the paper, when integrated with Max Plank's idea of the quanta, set the basis for the evolution of Quantum Mechanics. Einstein's second paper in the same year came out in May, where he assessed the age-old question of whether atoms exist or not. He showed that the behavior of particles randomly moving around in Brownian motion could be explained and predicted by the collisions of atoms, which was soon confirmed by experimentalists and thus, throwing all atom skeptics in the towel. His third paper, which was out in June, was the one which was finally able to set aside Einstein's 'psychic tension' (as he called it), which had been bothering him for decades. His paper toppled Newton's ideas of absolute space and time through the introduction of the Special Theory of Relativity. One might think at this point that Einstein ought to run out of ideas to produce something even more revolutionary, but in September, another paper arrived as a 'by the way this as well' kind of an article. Here, Einstein revealed that matter and energy were, in fact, equivalent and interchangeable as given by the equation $E = mc^2$ dubbed as the "*equation of the century.*"

The paper on Brownian motion was the most read and cited out of that year. Ironically though, it was not particularly groundbreaking and did not contribute much in front of the other three, which not only set the foundation of modern science but changed the way we understood the characteristics of light, space, time, mass, and energy.

After the submission of these papers, the academic world took notice of Einstein. Subsequently, he was offered numerous positions at different universities. He was appointed the leading scientist and a lecturer at the University of Bern in 1908 and promoted to Associate professor in 1909. In 1911, Einstein was appointed as a full professor at the Charles-Ferdinand University in Prague where he wrote 11 scientific papers mostly on mathematics and quantum theory of solids. Einstein returned to his alma matter at ETH Zurich to teach Mathematics and

Thermodynamics in 1912. In 1913, Einstein was also voted in as a member of the Prussian Academy of Sciences, which brought him back to Germany, his home country, in 1914. Moreover, after Einstein accepted his German citizenship back, he was elected the president of the German Physical Society in 1916 and later appointed as the first director of Kaiser Wilhelm Institute of Physics in 1917 shortly after its establishment.

Had Einstein retired back to the patent office after submitting those four ground-breaking papers in 1905 and achieved nothing else in his entire life, he would have still set the gold standard of a startling, unexpected genius. Nevertheless, Einstein was not finished. In 1915–16, he published yet another physics-breaking theory that dusted Newton's most famous work, gravity, off the cards. This was the General Theory of Relativity (sometimes referred to as Einstein's theory of gravity), which presented a broader and more comprehensive picture of how the universe works than his Special Theory of Relativity did. However, it was only later in 1919 when Einstein's general relativity was experimentally confirmed. That day was a momentous day and brought Einstein to world fame as the international media turned

LIGHTS ALL ASKEW IN THE HEAVENS

Men of Science More or Less Agog Over Results of Eclipse Observations.

EINSTEIN THEORY TRIUMPHS

Stars Not Where They Seemed or Were Calculated to be, but Nobody Need Worry.

Figure 1.2 Headlines of an article published in the New York Times about Einstein's theory in 1919.

him into a global icon. The New York Times said, *"Einstein Expounds His New Theory; Improves on Newton; Inspired as Newton but by the fall of a man from a roof instead of the fall of an apple."* Shown in figure 1.2 is an excerpt from an article published in the New York Times regarding Einstein's theory after it was experimentally confirmed in 1919 by the British expedition.

What were these groundbreaking ideas? Why are they so essential for us too? How do they contribute to our world? These questions naturally arise after learning about Einstein's influential impact. The reason to introduce his theories, especially to high school students, is that it gives an accurate account of the workings of the universe. After learning its so-called erudite concepts, one understands that although Newton ably explained the world and his ideas are extremely useful even today, they are undeniably outdated.

In further chapters, we will explore the theory of relativity (both the Special and the General Theory) and cosmology, which was born as a result of Einstein's General Theory of relativity. These physical theories may sound daunting at first because of the public opinion that they would naturally involve much-advanced mathematics and demand various pre-requisites from the reader. It is not true. All the concepts are introduced using creative thought experiments, and explanations provided for the conclusions are simple, relatively non-mathematical (mathematical equations are almost non-existent), seemingly unbelievable, but staggeringly convincing. The only pre-requisite is a yearning to learn the truth about the universe we call home.

CHAPTER 02

An Intro to the Special Theory

"An experiment is a question which science poses to nature, and a measurement is the recording of nature's answer."

– Max Planck

"No amount of experimentation can ever prove me right; a single experiment can prove me wrong."

– Albert Einstein

"It is reasonable to ask who or what created the universe, but if the answer is god, then the question has been merely deflected to that of who created God."

– Stephen Hawking

"Equipped with his five senses, man explores the universe around him and calls it science."

– Edwin Powell Hubble

We have familiarized the two fundamental principles behind the Special Theory of Relativity (discussed in the first section), quite simple, aren't they? Well, the simplicity of the foundations makes the theory itself quite simple, and that was one of the main reasons why

Einstein's Special Theory of Relativity was dismissed by the scientific community at first sight. It was just too simple! No advanced mathematics, no complicated derivations, no sophisticated hypothesis, or no expensive experimentation required. A couple of thought experiments, some grade nine mathematics, and a brain ready to challenge common sense, and a will to explore uncharted waters are all that took Einstein to topple our understanding of how the universe works for forever.

Einstein's Special Theory of relativity revolves majorly around two critical terms – 'relativity' and 'frame of reference' – and it becomes essential to comprehend their meanings in Physics. If you are well-versed with both these terms, you may skip the next two thought experiments to the sub-chapter *'Relativity in Danger.'*

Thought Experiment #1

Imagine: You wake up inside a room with no windows and one locked door (you checked). You cannot see outside. Looking around, you see a table with several items on it: a desk lamp (plugged in and turned on), a tennis ball, a ball of string, a pitcher of water and a cup, a candle, a box of matches, and a music player with headphones. The music player has a message on it that reads, *"Turn on for instructions,"* and so you do. As you listen to the recording, you learn that you are in a specially designed vibration-proof and noise-proof train car on a pair of straight and level train tracks. You are required to use one or more of the items in the room to determine whether the train car is at rest or is moving on the tracks. Destroying or modifying the walls, floor, or ceiling of the vehicle is not possible. Moreover, you are provided with a strict 30-minute time limit for the conduction of your test (or tests).

Note: For those who know about Foucault pendulums, note that the pendulum would only show whether the train car was rotating with the earth, not necessarily whether it was moving on the tracks. If the train were moving in such a way that its latitude position on the Earth was changing, a sensitive pendulum might detect the train's motion on the tracks. Nevertheless, for our thought experiment, we will assume that the train's latitude does not change.

Before you move forward, try to think of some creative ways to use the items available to you that might indicate whether you are moving or not.

Spoiler alert: The concept of relativity asserts that there is absolutely no way for you to tell whether the train car is moving or not.

A well-known fact is that one cannot describe the motion of an object without an external reference to it, a truth that cannot be falsified in this universe at least. Hence, *you will* eventually need to take assistance *from outside* to determine if the train is moving relative to earth or not (if you did find a way to conclude that you are moving or not, I suggest you re-analyze your method as there must be some loophole in it which you may have exploited to come to a conclusion).

The idea of relativity is that one can never define a state of **'absolute rest'** to any object, which means that no object in the universe can ever be stationary. To counter the statement, one may argue that in the given example, the earth is in a state of absolute rest, but that is not true. The earth revolves around the sun; hence, Earth can never be in a state of absolute rest, or stationary, with respect to the sun. In conclusion, every object is always moving with respect to some other object in the entire cosmos, and hence it is in relative motion. Here, it is vital to score a distinction between the term **rest** and **absolute rest** (stationary). An object can be at rest only with respect to another second object, but it can never be in an 'absolute state' of rest because it (the first object) will always be moving with respect to some other third object in the entire cosmos.

To understand the literal meaning of 'relativity,' consider a typical example. A car is moving across the road. Through a speedometer, I check the speed of the car and announce that it is traveling at 50 kmph. But, what do I mean by that? I mean that when I witness the motion of the car from my position, I will observe that it travels at a speed of 50 kilometers in one hour. Therefore, the car is speeding across Earth at 50 kmph relative to my position. My friend Alpha, who is sitting inside his car and moving at a speed of 20kmph, also sees that car and measures its speed with his speedometer. He announces that the car is traveling with a speed of 30 kmph. Because Alpha is himself moving at a speed of 20 and the other car at 50, he sees the other car moves at a speed of 30 by negating the effect of his motion. In this situation, both Alpha and I are right. Therefore, there is no absolute answer to what the speed of the car is, and we may only say that it is moving at 50 kmph 'relative' to my position and 30 kmph 'relative' to Alpha's position. Thus, one can specify the motion of an object only in relation to another object or a point – which is the literal meaning of 'relativity.'

For the physics nerd: In physics, the body where events, as described above, are spatially recorded is known as a coordinate system. The laws of Newtonian mechanics are always formulated on a coordinate system. Such a coordinate system, though, cannot be arbitrarily chosen. It obeys the laws of mechanics and is always free from rotation and acceleration, also called an 'inertial system.' One should remember that when any co-ordinate system is moved in uniform translatory motion (without rotation or acceleration) relative to an inertial system, that coordinate system is likewise also inertial. The first principle of the theory of relativity is the generalization of the statement above: every universal law of physics, which is valid in a coordinate system C, must also be valid in another co-ordinate system D that is in uniform translatory motion relative to C.

Thought Experiment #2

Imagine: You wake up inside a room which is completely covered from all sides except for a small window which has its shutter down. You look around and find a note which says, "You are inside a vibration-proof, noise-proof train with a window on one of the sides. When you pull up the shutters of the window, you will see another train outside that will be moving towards the right. Your job is to tell which of the trains are moving, yours or the other one."

(for the sake of this experiment, we will assume that the trains are extremely long).

Think it over and try to figure out the answer yourself.

In this case, there can be three answers – your train is moving the other is at rest, the other train is moving and you are at rest, or that both the trains are moving. But what is the answer? As it turns out, all three answers are correct. Inside the room, you do not feel any movement and conclude that it is the other train that is moving. So, in your 'frame of reference,' your conclusion is absolutely valid as you perceive the motion of the other train relative to your own. Although, in a similar room in the other train, your friend concludes that it is her train which is at rest and your train which is moving, and in her 'frame of reference' she is right. In a third scenario, your friend, in a helicopter up in the air, observes the trains moving past each other. In his frame of reference, your friend concludes that both the trains are running and undeniably, he too is right.

Therefore, the frame of reference is a setup through which an event can be *observed*. In different frames of reference, various conclusions can be drawn out, with all of them being correct in separate ways. Therefore, *there is no absolute frame of reference*, and every event occurs in relation (or relative) to another event.

Relativity in Danger

The concept of relativity, though, went through a rough patch before being firmly accepted. After the discovery of Maxwell's laws and the establishment that light is an electromagnetic wave, scientists were extremely sure of the presence of an aether-like substance as it was the only way to explain the transportation of electromagnetic waves across the cosmos. This belief, though, tended them towards believing that the idea of relativity is after all false because Maxwell's findings would hold only in the presence of aether, and if aether is present it will be omnipresent, giving it an absolute frame of reference, contrary to the idea of relativity. Moreover, the assumption of aether introduced another paradox. Any object moving in reference to this aether would not see the light traveling in all directions at the same speed because of drag and friction effects which will slow the light wave down, and this was again contrary to the results of Maxwell's equations which clearly predicted that light travels with the same speed in all directions irrespective of any external situation.

To solve this paradox, we must turn back a little. In the experiment for electromagnetic induction, we saw how, by moving a bar magnet in the vicinity of an electric coil, we could generate electricity. The theoretical explanation for this is that as the magnet moves near and away from the coil, the magnetic field changes and produces an electric field around the coil that produces an electric current. On the contrary, when we move the coil and keep the magnet at rest, we still generate electricity. The theoretical explanation for this phenomenon is given to be that if there is a moving charge (in this case, the wire) in the presence of a magnetic field, a force acts on the charge, and it starts moving around like a current.

When we look at both the cases as a neutral observer, we realize that both the instances represent the same physical situation, but we still end up giving two different explanations for it, very redundant indeed. Einstein, a firm believer in relativeness, found this extremely bothersome. He said that this just seems to violate the core idea of relativity, which has been widely known and accepted since the time of Galileo. He said that between the coil and the magnet, it should not matter whether the coil moves or the magnet moves because they both move relative to each other, and in each case, the magnetic field is responsible for producing the electric current. Through the idea of relativity, Einstein successfully explained the phenomena of electromagnetic induction by giving a single straightforward explanation.

Moreover, as mentioned earlier, scientists during the late 1800s were wildly looking for the aether. To this, Einstein argued that if there were such a thing called aether and it filled out the entire universe, then it would be at absolute rest because of its omnipresence and this again violates the idea behind relativity as there cannot be and must not be an absolute frame of reference. From this conjecture, he concluded that aether does not exist, and hence, electromagnetic waves must travel through a vacuum at the speed of light without the need for any external medium. Now, without the aether, we can define no particular reference frame in a vacuum. Thus, it must be concluded that the speed of light is independent of any frame of reference, and therefore, in simpler words, it is constant for everyone. Einstein's conclusion agreed with Maxwell's equations. Moreover, it also reinforced the idea of relativity which had been agreed upon for over three centuries since

Galileo studied balls rolling over inclined surfaces, or as the legend has it, by analyzing spheres falling from the leaning Tower of Pisa.

As it turns out, by the elimination of the idea of the aether, everything rightly fell into place. Results from Maxwell's equations well and truly coincided with the concept of relativity, which was also robustly complemented by the results of the Michelson-Morley **null experiment**. In the end, the idea of aether was indeed discarded based on concrete mathematical and experimental proof. More important, though, were the implications of the conclusions drawn by Einstein as the mathematical results gave bizarre results. It forced us to rethink how we usually measure length and time in the sense that they are not absolute. Just as motion is relative, through Einstein's Special theory of relativity, we will see that so are length and time.

Note: For the big fans of relativism, it is crucial to clarify that Einstein's relativity or relativity in general in physics has no relation whatsoever with relativism, which is an independent doctrine of philosophy.

CHAPTER 03

The Weirdness Begins

"We live in a society exquisitely dependent on science and technology, in which hardly anyone knows anything about science and technology."

– Carl Sagan

"Perfect as the wing of a bird maybe, it will never enable the bird to fly if unsupported by air. Facts are the air of science. Without them, a man of science can never rise."

– Ivan Pavlov

"Physics is the only profession in which prophecy is not only accurate but routine."

– Neil deGrasse Tyson

"When you are courting a nice girl an hour seems like a second. When you sit on a red-hot cinder, a second seems like an hour. That's relativity."

– Albert Einstein

The first significant implication of Einstein's principles was the conception of the idea that time is relative, which can be understood by the working of the light clock.

A light clock is a hypothetical clock solely imagined for the understanding of Einstein's Special Theory of Relativity. It consists of two mirrors placed parallel to each other. A light beam is flashed from one mirror that reflects back from the other mirror, and as soon as the light returns, a click is heard that signifies that one cyclic rotation of the light beam is complete. A qualitative analysis of light clocks can help one understand the concept of time working differently for people in different frames of reference. Two light clocks are given, one each, to Mahesh and Geeta. The distance between the mirrors of both the clocks is equal and denoted by 'L.'

Initially, both Geeta and Mahesh are at rest with respect to each other, shown in figure 3.1. To check that their clocks are properly synchronized and working, we assume that both our participants have two atomic clocks, each which measures the time of one rotation of the light beam inside the light clocks. After Geeta and Mahesh start their light clocks and measure the time between on rotation, they observe that the light clocks are adequately synchronized as both the atomic clocks display the same time for the light beams, which travel distance 2L, to complete one rotation. Moreover, they hear a cohesive sound of clicks after each rotation of light is completed; this further confirms that the light clocks are working correctly.

Things become interesting when Geeta and her clock are set in motion. From Geeta's frame of reference, she and her light clock are stationary with respect to each other, and when she measures the time of one rotation for the light beam, which travels a distance 2L as shown previously, her atomic clock shows the same time as it did initially, hence no issues.

The Weirdness Begins

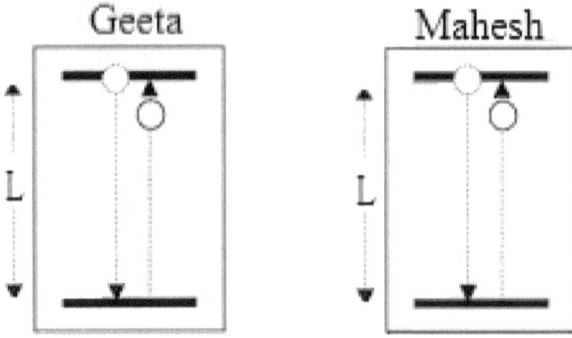

Figure 3.1 Geeta and Mahesh's light clocks when stationary. Both the light clocks are at rest in relation to each other and hence, will remain in sync.

Although, something weird happens if Mahesh looks at Geeta's clock from his frame of reference. According to Mahesh, Geeta is moving with her clock at a certain speed in the forward direction. Therefore, he sees the rotation of the clock's light beam in a three-step process (as shown in figure 3.2).

When Geeta starts her clock, Mahesh sees that the light leaves mirror A in position one. As the light travels towards the mirror B, Mahesh sees that the mirrors are no longer in position one but have moved to position two; so, for the light to reach the mirror B, it has to travel diagonally, a distance 'D,' to contact mirror B in position 2. Similarly, when the light reflects back from mirror B, in position two, Mahesh sees that the light reaches mirror A in position three because the mirror moves forward while the light is traveling between the mirrors. So, Mahesh sees that the light travels a distance 2D instead of 2L, and, by simple Math, he says that 2D is greater than 2L.

Putting it mathematically, the speed of light is the same in both the situations {principle of light constancy}, whereas the distance traveled by light is higher in Mahesh's frame compared to that in Geeta's frame. Thus, the time taken by light to move up and down will be higher in Mahesh's frame compared to that in Geeta's frame. The only logical conclusion which explains the contrast in the two situations is that 'time runs slowly for Geeta.' Generalizing the outcome, we can say that *time runs slower for moving objects than it does for objects at rest*. This generalization, though, is a little ambiguous to understand because one cannot announce in absolute terms that a particular object is moving while the other remains at rest {principle of relativity}. Let us understand this contradiction further.

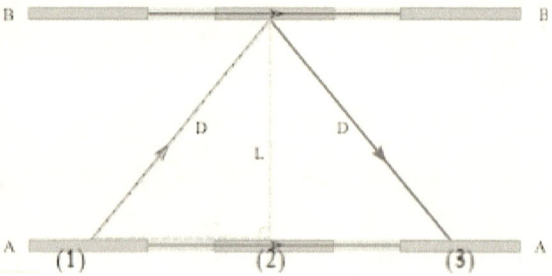

Figure 3.2 Geeta's clock in motion as seen by Mahesh from his frame of reference

In the above thought experiment, we analyzed the situation from Mahesh's perspective and concluded that the time for Geeta is running slower than it is running for Mahesh. If we consider the same situation from Geeta's frame of reference, Geeta announces that she is at rest, but it is Mahesh who is moving backward. So, after conducting the same analysis for Geeta, which was done for Mahesh, we observe that it is now Geeta, who sees time running slower for Mahesh when compared to hers. Hence, it must be concluded that similar to motion, *time is also relative*, and depending upon the frame of reference, time runs slower for the object that is moving compared to time running for the object that is at rest. In other words, each observer at different positions in space, traveling at different speeds, needs to have his/her measure of time, because the measure of time as calculated by their clocks may not agree with the measure of time calculated by identical clocks carried by other observers. This phenomenon is also known as the *time dilation* effect.

While stating that time is running slower for a particular object which is moving with respect to a second object relatively at rest, I literally mean that any clock, from an atomic clock to a bedroom clock, will run slowly for that moving object. Really! This statement is contrary to what we experience in our day-to-day life because effectively speaking, we do not observe clocks running slower if we travel in a car, train, or even an airplane. The reason why we do not feel the slowness of time is that the effect is minimal for us to notice and becomes significant only at speeds nearing the speed of light. To put the situation into perspective, understand this – an average airplane flies at approximately 0.00000083 times the speed of light, which, after calculation, means that time slows down by a factor of 1.00000000068 times for the airplane, almost negligible.

To understand the effects of the slowness of time, we must also familiarize ourselves with the '**Lorentz factor**' (sometimes also referred to as the 'gamma factor' or only 'gamma'). The Lorentz factor is a simple mathematical term that is integral to the special theory of relativity as it explains the weird phenomena that are predicted by the theory in a straightforward method.

Note: The Lorentz factor further gives rise to Lorentz transformations, whose mathematical understanding comes from the concept of Galilean transformations, which to avoid complex mathematics, have not been discussed in the book. The Lorentz transformations are introduced in subsequent sections for a qualitative understanding with minimal mathematical terms to avoid complexity. Those who might want to refer to the derivations of the Lorentz factor and Lorentz transformations may refer to their derivations on physics.stackexchange, Khan academy, or the Byju's app. The first website mentioned provides the most straightforward explanation, while the other two are a little tedious and unnecessarily complicated.

The Lorentz factor is (represented by gamma): $\gamma = \dfrac{1}{\sqrt{1 - \dfrac{v^2}{c^2}}}$

Here 'v' represents the speed of the moving object and 'c' represents the speed of light, which is always constant. Given below are some cases which help us to get an idea of how the value of γ changes according to 'v.'

Speed in km/s (v)	The Lorentz factor (γ)
0.5	1.0000000000014
300	1.0000005
30,000	1.005
150,000	1.2
297,000	7.1
299,970	71

As one can see, the gamma factor starts to rise dramatically as the values of 'v' are initiated towards the speed of light. An interesting point to consider here is that the gamma factor will always remain higher than or equal to one, no matter what value 'v' is given between 0 and 300,000. We will discuss what happens when we put 300,000 in later chapters. (There is no need to remember these values as these have been presented only for illustration)

Relativity of Simultaneity

Another significant implication of Einstein's principles is the relativity of simultaneity. The concept is that the possibility of distant simultaneity – two spatially separated events occurring simultaneously – is not absolute, but depends upon the observer's reference frame, or *'time is suspect,'* as Einstein put it. To understand this concept, it is essential to exercise the mind extensively through thought experiments. So, start imagining.

Geeta, our good old friend, is on a spaceship in outer space. In the exact middle of her spaceship is a laser gun that can shoot lasers to the front and the rear of the ship at light speed 'c.' Similarly, Mahesh is on another spaceship, which is equal in length to Geeta's ship in outer space next to her. In the middle of Mahesh's spaceship is also a laser gun that shoots lasers across the ship to the front and the rear. Now, assume that both Geeta and Mahesh are at rest with respect to each other. Both of them keep two clocks each, one at the front and the other at the rear, in their respective ships, with both of the clocks being synchronized to show time zero. The experiment starts. Both, Geeta and Mahesh, shoot lasers from the gun to the front and the rear of the ships simultaneously at time t=0.

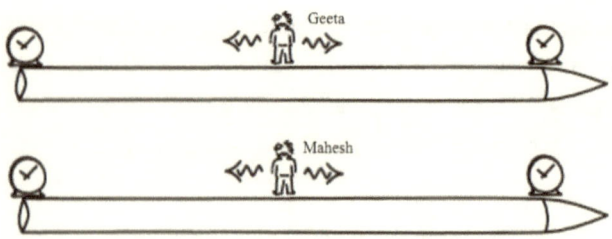

As soon as the lasers reach the ends of the ships, the clocks stop ticking and record the time of contact of the laser. Suppose Geeta marks the time on her clocks as 'Ta' while Mahesh marks the time in his clocks as 'Tb.' In Mahesh's frame of reference, Geeta is at rest, and he agrees with Geeta that both lasers reach the ends of her ship simultaneously at 'Ta.' In Geeta's frame of reference, Mahesh is at rest, and she agrees that both lasers reach the ends of his ship simultaneously at 'Tb,' and because the lengths of the ships are the same, we conclude that Ta=Tb. No complications, no questions, and we proceed further.

In a second case, Geeta's ship is moving forward at a velocity 'v' with respect to Mahesh's ship. Both of our trusted friends synchronize their clocks at time t=0 again and repeat the same experiment as both of them shoot their respective lasers. Consider what happens in Geeta's and Mahesh's frame of reference separately.

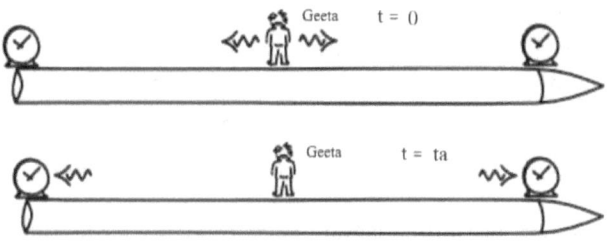

In Geeta's frame of reference, she sees that her spaceship is at rest as nothing moves inside the ship. So, as both the lasers strike the ends of her spaceship and the clocks stop ticking precisely at the point of contact, she compares the time in both the clocks and notes that the lasers reach their respective ends simultaneously at time t=Ta, and undoubtedly in her frame of reference, Geeta is correct.

Now, when Geeta sees Mahesh's ship in her frame of reference, she assumes that it is her ship which is at rest and quiet clearly it is Mahesh's ship which is moving backward with velocity 'v.' So, when Mahesh shoots the lasers across his ship, she sees that the laser moving to the front of his ship has to travel less distance as the ship moves opposite to the direction of that laser at speed 'v' and covers, let's say, a distance 'D' before the laser strikes the end. So according to her, she records the time of contact of the laser traveling to the front of Mahesh's ship as 'Tb1'.

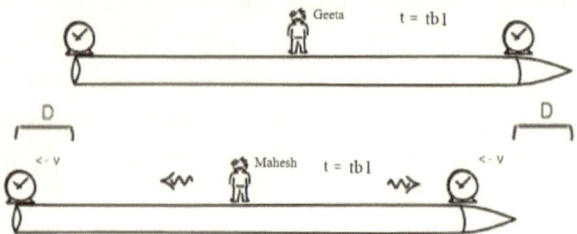

At Tb1, the laser traveling to the rear still has not reached, and when it does arrive there, Geeta records the time of contact with the end as 'Tb2' and the let us say the distance traveled by ship until that point in time be 'd.'

As we see in the diagrams, according to Geeta, the laser in front reaches the clock at a time 't=Tb1' whereas the laser to the rear reaches the dropping at a time 't=Tb2'. In her frame of reference, Geeta is correct.

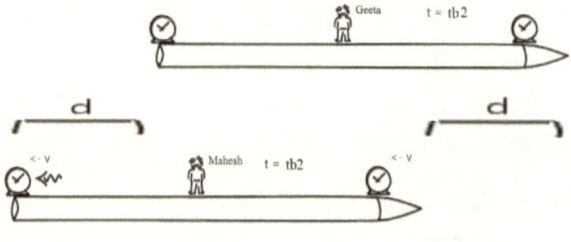

Important Note: One may argue that the speed of light as Geeta witnesses it in Mahesh's ship is 'c-v' for the rear and 'c+v' for the front lasers considering that the light and the ship's velocity will add up according to the *'vector law of addition.'*. This conclusion, although, is not possible because of the *'principle of light constancy'* which states that the speed of light for any observer moving at any speed will always be 'c.' So, irrespective of whether the laser travels to the rear or the front of the ship, Geeta still sees the lasers traveling at the speed of light, and the same is applicable for Mahesh in his frame of reference.

Considering Mahesh's frame of reference, though, he sees the events turn out differently in Geeta's spaceship. According to him, it is Geeta who is moving forward at a velocity 'v.'

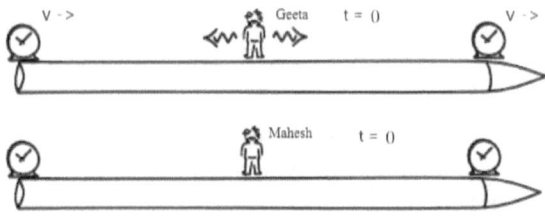

Therefore, in his ship, Mahesh sees nothing move with respect to each other, and hence, his ship is at rest. So, after he shoots the lasers and both of them strike the ends of his spaceship, the clocks stop ticking precisely at the point of contact and he compares the time in both the clocks. He clearly notes that the lasers reach their respective ends simultaneously at time t=Tb, and undoubtedly, in his frame of reference, Mahesh is correct.

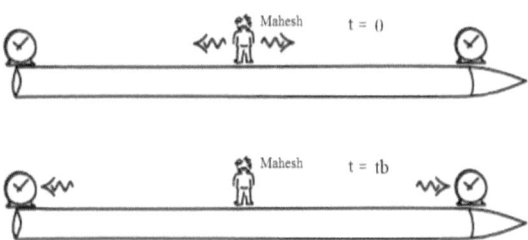

However, when he sees Geeta's ship, because, in his frame of reference, she is moving forward with velocity 'v,' he sees things turn out differently than Geeta does. He witnesses that the laser going to the rear of Geeta's ship has to travel less distance than the laser moving to the front because the ship travels a distance 'D' due to its velocity 'v' in the forward direction. So, Mahesh sees that the laser reaches the rear of Geeta's ship sooner at t=Ta1.

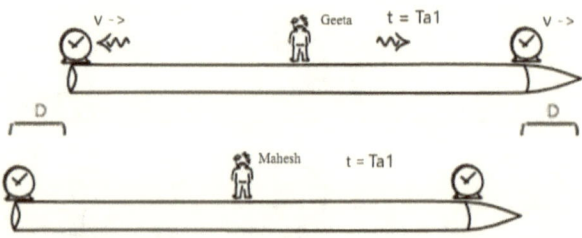

Furthermore, because the laser traveling to the front of the ship has to travel a longer distance as the ship is moving away, he records the time of contact of the laser with the front of the ship at t=Ta2 as the ship covers a total distance 'd.'

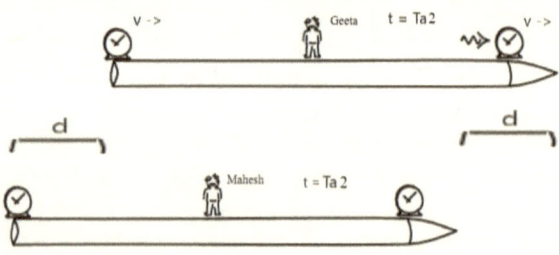

Let us simplify the results by summarizing the conclusions derived from the thought experiment for better understanding.

Geeta's conclusions (in her frame of reference) –

1. The lasers reach the ends of her spaceship simultaneously in time = Ta.

2. The lasers do not reach the ends of Mahesh's ship simultaneously as the laser traveling to the front reaches before the one traveling to the rear.

Mahesh's conclusions (in his frame of reference) –

1. The lasers reach the ends of his spaceship simultaneously in time t = Tb.

2. The lasers do not reach the ends of Geeta's ship simultaneously as the laser traveling to the rear reaches before the one traveling to the front.

After comparing their conclusions, Geeta tells Mahesh, "Mahesh, you have quite clearly screwed up your clocks; they are not synchronized properly as my clocks clearly show that the lasers in your ship do not reach simultaneously." To this, Mahesh replies, "No. It is your clocks which are messed up as my clocks show that my lasers reach their respective ends of the ship simultaneously, but it is in your ship where they do not reach simultaneously and your observations are flawed". An altercation breaks down between Geeta and Mahesh as they argue that it is the other person's observations that are at fault. Though, after analyzing the complete scenario and according to their respective frames of reference, both Geeta's and Mahesh's observations are correct. But, how does one explain the discrepancy which arises between the two observations?

We know that Geeta is moving forward with a velocity v while Mahesh observes her from his ship, which is at rest. In Mahesh's frame of reference, he observes that the laser going to the rear end of her ship has to travel a lesser distance than the laser moving to the front end because of the motion of the ship itself. Geeta, in her ship, takes a photograph of the clocks at the instant when both the light beams strike their respective ends and, on comparing, both her clocks show the same time, agreed. This value of time is lesser than what Mahesh should ideally observe, which can be explained because of the time dilation effect. However, what cannot be described here is that Mahesh observes the two light beams strike the ends of Geeta's ship at different time intervals!

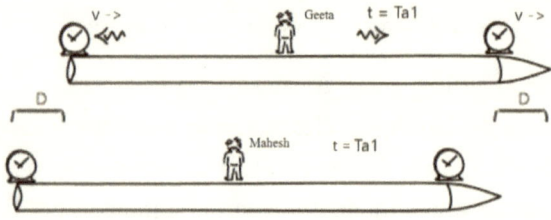

To consider this, we need to understand the concept of events. An event is a specific manifestation of an action in space which takes place at one particular point in time, but it may be observed by different observers moving at different speeds in different ways {relativity of simultaneity}. In this situation, it is essential to take note of the fact that the event of the light beam just contacting the ends of the spaceship happens only once and that too at the same point in time, according to Geeta. Now, because the light beams reach at the same time for Geeta, they cannot arrive at different points of time for Mahesh because that will turn into a separate event, whereas we know that we are dealing with the same event here. So, in Mahesh's frame of reference, although he observes that the light beams in Geeta's ship do not reach the ends of her ship simultaneously, he must still observe them reach the ends of her ship at the same point in time. For this to happen, Mahesh must observe that the clock in the front of Geeta's ship runs slower than the clock in the rear of her ship. Thus, if the light beam reaches the rear end at say time t=Tb1 for Mahesh, the light beam traveling to the front end must also reach at time t=Tb1.

Hence, it should be concluded that time runs slower in the front end of Geeta's spaceship than it runs at the rear end of her ship. Therefore, because, according to Mahesh, the clock in the front of Geeta's ship is running slower than the one in the rear end, he observes that the light beam reaches the rear end of Geeta's ship before it reaches the front end. This phenomenon of time running relative to each other at different ends of an object is known as the *leading clocks lag effect*. Primarily, this effect means that time runs slower in the front of a moving object when compared to the rear of the same moving object.

Therefore we can conclude that because of the relativity of simultaneity, Mahesh observes that the event of light beams striking the ends of Geeta's ship does not occur simultaneously in his frame of reference, but because of leading clocks lag factor, Mahesh also observes that the light beams eventually strike the ends of the ship at the same time to which, Geeta also agrees. So, Tb1 and Tb2 are necessarily equal in the above thought experiment.

For the physics nerd: The mathematical relation for the time lag in the leading clock (clock in the front end of an object) or the amount of time by which the front of a moving object lag by the rear of the object is given by $T = \dfrac{Dv}{c^2}$. Where T is the time lag between the two ends of the moving object, D is the *'proper length'* of the object, v is the speed of the object, and c is the speed of light.

Here, it is vital to take note that the 'time dilation' effect and the 'leading clocks lag' effect are both different and not inter-related in any way. Due to the time dilation effect, time runs slowly for Geeta in Mahesh's frame of reference, which is true and well established. But, due to the leading clocks lag effect, time runs slower in the front of Geeta's ship when compared to time running in the back of her ship, as observed from Mahesh's frame of reference, which is not only accurate but a well-established phenomenon as well.

To completely comprehend and imagine both these effects running in tandem is a nightmare for physicists as each time one imagines the scenario, the reality completely defies logic, and thankfully for us, Einstein recognized that. We will discuss the working of both these effects while studying the Pole in the Barn paradox in the chapter *'Paradoxes to Ponder,'* which will clear things out to a greater extent and provide a better understanding of both the effects working in tandem as well as separately.

CHAPTER 04

Playing with Length & Time

"A scientific truth does not triumph by convincing its opponents and making them see the light, but rather because its opponents eventually die and a new generation grows up that is familiar with it.

– Max Planck

"If you wish to make an apple pie from scratch, you must first invent the universe."

– Carl Sagan

"To raise new questions, new possibilities, to regard old problems with a new angle, requires creative imagination and marks real progress in science."

– Albert Einstein

"When radium was discovered, no one knew that it would prove useful in hospitals. The work was one of pure science. And this is proof that scientific work must not be considered from the point of view of the direct usefulness of it."

– Marie Curie

So far, by doing a qualitative analysis of Einstein's ideas, we understood the concept of time dilation, which states that time, after all, is relative and runs slower for a moving object with respect to a frame of reference at rest. A quantitative analysis is required to explore the practical application of the concept of time dilation.

After performing necessary mathematical calculations, a relation between the time variants of two different frames of references was established. This relation is in terms of the speed of time and is given by a simple formula:

$$\Delta t\ (\text{observer}) = \frac{1}{\gamma} * \Delta t\ (\text{actual})$$

The triangle-shaped sign is called delta (a Greek symbol) and is generally used to represent a change in a particular quantity over a specified period. Here, the sign represents the change in time for the moving object in terms of the change in time for the object at rest. As previously discussed, γ will always remain greater than one. Hence this equation signifies that any clock, in general time, will always run slower for an object which is moving with respect to the time running for the same object when at rest.

Consider a situation where Geeta is moving in a spaceship with respect to Mahesh, who is at rest, at 0.99995 times the speed of light, i.e., approximately 299,985 km/s. The Lorentz factor for this velocity is 100. After doing the math, it is found that time will run 100 times slower for Geeta than it will for Mahesh. To put this into perspective, presume that both Mahesh and Geeta are 20 years old when Geeta leaves in her spaceship at speed mentioned above. Mahesh, somewhat improbably and surprisingly, lives another 100 years till he is 120 when Geeta returns. Now, because time is running 100 times slower for Geeta, according to her, only one year has passed during the entire trip, and she expects to see Mahesh 21 years of age, as young as herself. However, when they meet on Geeta's return, Geeta realizes that although she may be 21, Mahesh has aged an entire century because of the effect of time dilation. (It is presently humanly impossible to achieve Geeta's speed, but only for the sake of our understanding we assume that it is happening.)

Therefore, if Geeta were to travel with that speed in reality, things will turn out the way they did in the thought experiment, which is universally agreed upon, but Geeta does not agree. According to her, she is the one at rest, and it is Mahesh who moves with the entire Earth in the backward direction at the speed of 0.99995C ('C' represents the speed of light). For her, it is Mahesh's clock that should run 100 times slower than hers, and in her frame of reference, she is correct in assuming that. This point of view is although contrary to what happens in reality and thereby introduces a contradiction. To untangle ourselves from this mess, we must now learn about the relativistic effects of length.

Length Contraction

The concept of length contract is a necessary consequence that arises from time dilation. To understand it, we keep aside the example of Geeta and Mahesh we discussed above and will turn back to it at the end of this section.

To explore this idea of length contraction, our friends Mahesh and Geeta are required to measure the length of one of their spaceships (assuming both are equally long) with a condition – one of the two experimentalists will be in one motion while the other remains at rest. So, Mahesh decides to remain at rest (somewhere in outer space) while Geeta passes by him in her spaceship at a velocity 'v.' Now, Mahesh has two clocks at the ends of his ship and Geeta has two clocks at respective ends of her ship, perfectly synchronized, which are used to measure her ship's length.

When the front end of Geeta's ship just passes by the rear of Mahesh's ship, as shown in the figure, a photograph is taken with both, Geeta's and Mahesh's, clocks in the frame. At this point, Mahesh records time Tb1 in his clock and Geeta records time Ta1 in her clock.

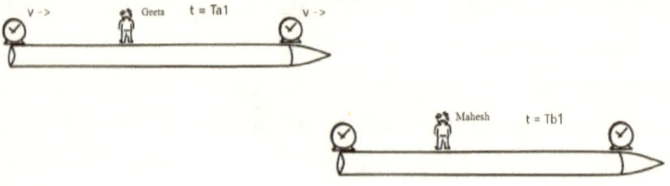

As Geeta's ship moves forward and the rear end of her ship exactly passes by the rear end of Mahesh's ship, another photograph of Geeta's and Mahesh's clocks is taken in a single frame. At this point of time, Mahesh records the time in his clock as Tb2 and Geeta records the time in her clock as Ta2.

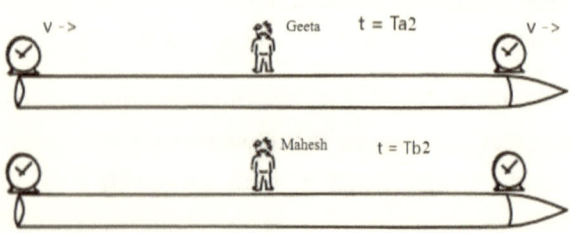

Playing with Length & Time

Importantly, because of photographic proof of both the clocks at the instant of exact crossing over of the ship's, the observations are authentic and indisputable.

After recording the readings of their respective clocks, Geeta and Mahesh get together to analyze them. To measure the length of Geeta's ship they make use of the formula, *'Length of object = Speed of object * Time taken by light to travel across.'* Now, in Mahesh's frame of reference, he recorded the time variance between the passing of two ends of Geeta's ship as 'Tb2 – Tb1' and length of Geeta's ship according to him will be La = v * (Tb2 – Tb1). In Geeta's frame of reference, the time difference between the passing of two ends of her ship will be 'Ta2 – Ta1', and hence, the length of her ship, according to her, will be: La = v * (Ta2 – Ta1).

After putting in the values, Geeta and Mahesh find that their answers do not match. Why does that happen? Clearly, the speed of Geeta's ship is the same for both our experimenters and so should the time difference be. However, that is not true because of the effects of time dilation. It is known that time runs slower for the moving object relative to an object at rest, which in this case means that Geeta's clock runs slower in comparison to Mahesh's clock. Hence, her recorded time interval will be shorter than that of Mahesh's, and this difference in readings produces an anomaly in the lengths calculated.

After doing necessary mathematical calculations, to quantitatively measure the amount of length contraction, a relation between the lengths of the same particular object in a moving frame versus in a frame of rest was established. The relationship is given by the simple formula:

$$\Delta L\left(\text{observer}\right) = \frac{1}{\gamma} * \Delta L\left(\text{actual}\right)$$

The above equation is very similar to the equation of time dilation, and hence, so are the results. By evaluating this equation, it can be concluded that moving lengths contract. Therefore, for moving objects, time runs slower and they themselves contract in length. So, in the above example, Mahesh measures Geeta's ship to be shorter than it actually is according to Geeta because of length contraction. Though for Geeta, she observes things differently. In her frame of reference, it is Mahesh's ship that is shorter than hers, and it is again Mahesh's ship where time runs slower than it does in her ship.

Note: Sometimes, the term 'proper length' is used to refer to the length of the object when it is measured at rest. Einstein used the term *'own length'* for that purpose.

After studying the concept of length contraction, a question naturally arises, why don't we feel the effects of length contraction in our day-to-day life?

The first point to consider is that just as time dilation becomes significant only at near light speeds, length contraction also becomes substantial only at near-light speeds. And, even though, if we somehow do make it close enough to light speeds, we will still practically not be able to feel the effects of length contraction. To understand that, consider the second point. Assume that Mahesh is traveling at a speed where the gamma factor is 2. So, technically speaking, the length of his spaceship should become half of its original value, but he does not observe that. According to Mahesh, everything inside the ship reduces by a factor of half and if he takes a meter scale and measures the length of a wood piece (whose length is known to be one meter), he will still measure it as one meter because both the meter scale and the wood piece together reduce by a factor of half. Another argument to prove this is that Mahesh considers himself at rest, and according to him, it is the universe outside which is moving backward. Hence, for Mahesh, it is the universe which has contracted by a factor of two, while he and his spaceship remain unaffected.

Coming back to the situation discussed in the previous section, we asked the question that according to Geeta, it should be Mahesh's clock which should run 100 times slower. So, according to her, Mahesh should not have aged, but that is contrary to what Geeta observes. In this situation, Geeta forgets the effect of length contraction. In her frame of reference, when she observes Mahesh fly backward at the speed of 0.99995C, she also observes that the universe contracts 100 times shorter {length contraction}. Therefore, in her frame of reference, Geeta observes that the distance Mahesh travels in 100 years for her took one year to travel because the universe itself contracted 100 times. In conclusion, Mahesh agrees that because time was running slower for Geeta, he aged a century while she grew only one year old, and Geeta agrees that because the universe contracted for her, she traveled the same distance in one year that Mahesh would have taken to travel in an entire century which is why he aged a whole century.

This is still only a rough idea to explain the situation at hand. There are other factors at play which we will discuss at length in a similar kind of situation in the chapter *'Paradoxes to Ponder.'*

Einstein and Special Relativity

It is interesting to note here that although Einstein did come up with these ideas (length contraction and time dilation) individually between 1902 and 1905, he was not the only one to produce their mathematical equations. Scientists like Hendrik Lorentz and Henri Poincare also came up with similar results, but they did so in a very different fashion. While studying Maxwell's equations in electrodynamics, Lorentz and Poincare were trying to figure out the nature of aether when the results that we have studied so far and many others quite inexplicably popped out of their work but also perfectly fit their analysis. They could neither make much sense of the equations nor could they get rid of them in the world of electrodynamics. It was Einstein who came up with these results by integrating two fundamental principles – the principle of relativity and the principle of light constancy – to give context to the weird results which Lorentz and Poincare did come up with separately but could not adequately explain. Lorentz factor, which we have so commonly used in the Special Theory of Relativity, was one such equation that popped out from nowhere in Lorentz's works of electrodynamics.

After the establishment of Einstein's Special Theory of Relativity, it was accepted that time is not an independent quantity but is combined with space itself. This concept gave rise to the idea of space-time. Generally, to describe the position of a specific event, one uses two coordinates or even three, if the event occurs in space or at some height, to do so. Though, now it becomes essential to give a fourth coordinate of time as well because time is relative and not the same for each observer.

In special relativity, there is no real distinction between space and time, and the four coordinates which are used to describe the position of an event are represented on a 4-dimensional plane called space-time graph. This four-dimensional plane is tough to visualize, given that it is not encountered commonly in the surroundings as the eyes naturally operate in a way to identify the world in only 3-dimensions. In fact, the discovery of the presence of more than three dimensions is an excellent achievement of scientific pursuit. Even greats like Stephen Hawking have admitted finding it extremely difficult to visualize even a simple graph in three-dimensions let aside the entire cosmos.

Personally, the most convincing way I have found to understand the nature of 4-dimensional space is by looking at a simulation of the tesseract. At this moment, a Marvel fan, quite understandably so, would be surprised and probably announce, "but hey! The Tesseract is the cube that contains the space stone" and so it is, but only in the comics. In reality, a tesseract is a 4-dimensional cube.

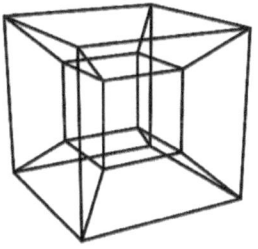

Figure 4.1 A tesseract is a cube in 4-dimensions.

Relativity through Graphs

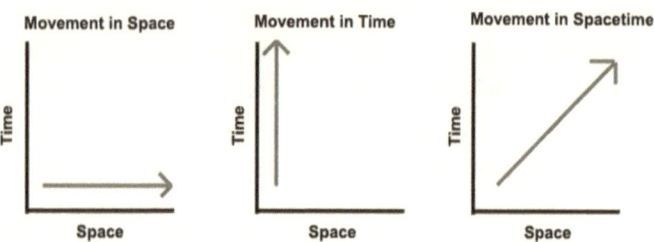

Figure 4.2 Three spacetime graphs depicting the movement of an object through only space, through only time, through both space and time (at an equal rate), in the respective order as shown above.

In physics, graphical methods often prove to be game-changers by providing us profound insights into concepts that we otherwise find challenging to comprehend. One such type of graph is the spacetime graph. It helps us to visualize the motion of an object by plotting that motion with separate coordinates of space and time on the axes of space and time combined. In a 2-dimensional spacetime graph (as shown), only the one-dimensional motion of the object is considered through space, assuming that it travels in a straight line (this makes life easier for the physicist due to less complication).

Let us conduct another thought experiment to witness the usefulness of spacetime graphs.

Monal, Mahesh's cousin, is speeding on a forest track on his skateboard at 60 miles an hour in precisely the North direction (vertically upwards as shown in figure 4.3 (a)). After some time, Monal climbs on a highway heading towards North-East direction. Now on the highway, Monal is moving at a speed of 60 miles an hour in the North-East direction (diagonally moving towards the top right corner on a graph as shown in figure 4.3 (b)). Through trigonometry, one can quickly conclude that Monal's speed will get equally divided on both co-ordinate axes – North and East - when considered separately.

So, an observe heading solely in the North direction will notice that Monal is not moving at 60 miles per hour but at $30\sqrt{2}$ miles per hour in the Northern direction. Similar will be the observations made by an observer heading in the Eastern direction. Hence, both observers would primarily observe that Monal is moving slower from their respective frames of reference ($30\sqrt{2}$ miles per hour) than he actually is (60 miles per hour).

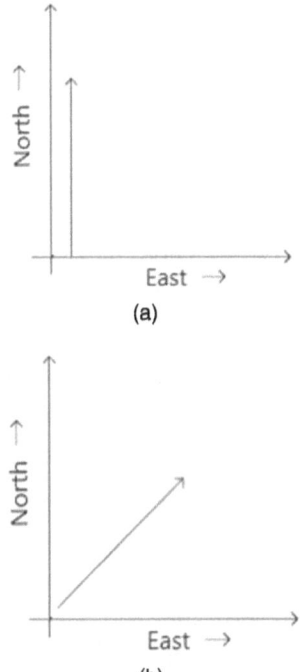

Figure 4.3 (a) Monal going in North direction. **(b)** Monal going in North-East direction.

Similarly, let us consider another situation in which a car is standing outside a restaurant. Now while the car is standing, it is essential to note that it does not move through space, but it still moves through time, which constantly ticks for it. This motion of the car can be depicted on a spacetime graph (as shown in figure 4.4 (a)), and a vertical line parallel to the y-axis can consequently be obtained. When this car sometime later moves on a road, it moves through space. Graphically depicting this motion, the line on the graph (shown in figure 4.4 (b)) slightly tilts and moves in a diagonal direction. When this happens, the motion of the car gets equally divided into space and time (similar to how it got split in Monal's case). When this happens, observers who are relatively at rest observe that time runs slower than usual for the car, and its length also contracts (similar to how the observers observed Monal's speed to reduce in the previous case).

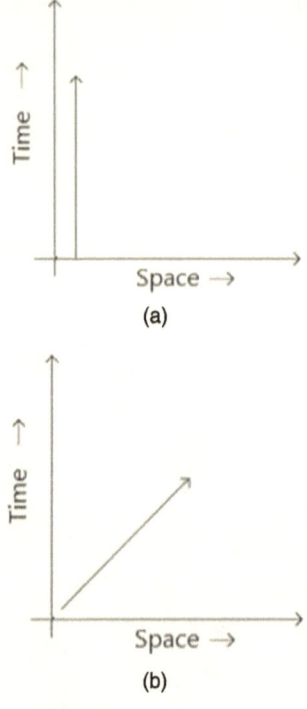

Figure 4.4 (a) Car travelling through time. (b) Car travelling through both space and time

This analogy is beneficial in visualizing how time and length actually work in reality.

CHAPTER 05

Paradoxes to Ponder

"The scientific theory I like best is that the rings of Saturn are composed entirely of lost airline luggage."

– Mark Russell

"He hoped and prayed that there wasn't an afterlife. Then he realized that there was a contradiction involved here and merely hoped that there wasn't an afterlife."

– Douglas Adams

"Life is a preparation for the future, and the best preparation for the future is to live as if there were none."

– Albert Einstein

"If you try to fail, and succeed, which have you done?"

– George Carlin

Einstein's theory of relativity, as you may have realized till now, absolutely defies general opinion. The ideas and concepts are too abstract and radical because although such events do occur in real life, we just do not seem to observe or recognize them; this is the reason why the theory ends up with a bucket full of paradoxes. Interestingly enough, though, no matter how assertive the statements of the paradox seem at first sight, after genuinely applying the theory of relativity, one always ends up concluding that the paradox is after all not too paradoxical but well explained.

In this book, we will talk about three significant paradoxes. While unearthing the real eventual happenings in these paradoxes, an aggressive in-depth analysis of the theory is required. Essentially that means that we need to explore a little bit of Physics and employ some minor mathematical analysis to understand what is going on and how to explain the situation. This is, in fact, the most gripping part of the entire theory. It is here that we actually challenge the theoretical concepts to understand their effects and how they are applied practically, which is hugely stimulating. But, even still, if you are not a big fan of a little bit of math, some complex analysis, and insanely logic-defying but incredibly enthralling reality, I would suggest that you skip to the next chapter of *'Special Theory in Real Life.'*

An important fact to consider while working within the confines of the Special theory of relativity is that we assume there are no effects of inertia, gravity, or any other force which might introduce acceleration or deceleration in the situation. So, if a spaceship is said to be traveling at a speed of say 0.6*C, in special relativity, it is assumed that the ship starts at the exact speed of 0.6*C, regularly travels with it throughout the journey, and instantly comes to rest when required.

Bell's Spaceship Paradox

This is a relatively simple and one of the easier paradoxes to solve out of all the paradoxes in the theory. In this paradox, we have an observer A and two spaceships B and C, which are tied to a rope that is a length 'D' apart. The spaceships are initially at rest and after every ten seconds, the observer A gives both the ships a signal which directs them to accelerate forward and increase their speed by a certain amount 'v' each time. The situation has been depicted in figure 5.1.

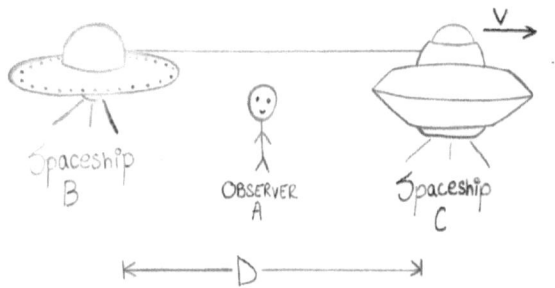

Figure 5.1 Representative image for the initial situation in Bell's spaceship paradox

One may say that we do not deal with accelerations in Special theory, and that is correct. In this scenario, to understand the concept of the Special theory, we will neglect the effects of General theory, which is where we deal with accelerations.

The conditions here are that, firstly, when spaceships B and C accelerate at the same time, they do so while keeping in mind that the length D of the rope remains constant for observer A, and secondly, they accelerate only after

10 seconds have passed for the observer A. The question now is, does the rope break? Looking at the situation from a classical point of view, we can say that as the distance D remains constant, which is a pre-requisite for the experiment in the first place, the rope should clearly never break. The scientist J S Bell, though, tells us otherwise.

When the observer gives the signal for acceleration, both the spaceships obediently accelerate at the same rate while keeping the distance of the rope constant and achieve a speed 'v.' Now, the observer also signals that after ten seconds both spaceships should accelerate once again. At this point, the spaceships have already accelerated once and are moving forward at a constant velocity v. From observer A's point of view now, he considers the system of both the spaceships and the rope to be one single object which should obey the leading clocks lag effect, and according to that effect, the observer A concludes that time runs slower for spaceship C, which is in the front (it is leading) than it does for spaceship B, which is in the rear.

So, let us say that after ten seconds pass for spaceship B, only eight seconds go off for spaceship C in observer A's frame of reference. Therefore, when spaceship B starts to accelerate, spaceship C does not accelerate because, as observer A observes, the ship still has two seconds to go before he signals it to accelerate. However, now, because spaceship B has accelerated and spaceship C has not, spaceship B comes closer to spaceship C and the distance of the rope between them is reduced. This violates the condition stated before which says that the length of the rope should always remain constant 'D.' So, in the observer A's frame of reference, to obey the condition, spaceship C should accelerate at eight seconds while spaceship B should accelerate at ten seconds so that the distance 'D' between the ships does not reduce, and this is precisely what he decides.

Now, in the spaceships' frame of reference, time runs at the same speed for them because as they see the situation, it is they who are at rest but it is the observer A who is moving backward at a velocity 'v.' So, when the spaceship C accelerates after eight seconds, the spaceship B still doesn't accelerate because in its frame of reference, it observes that only eight seconds have passed and the observer A, according to the new instructions, has clearly instructed it to accelerate only when ten seconds have passed for it. Therefore, when spaceship C accelerates, spaceship B does not accelerate, and from their frame of reference, the distance between the ships increases, and consequently, the rope attached between them breaks apart.

From observer A's frame of reference, though, the spaceships accelerate simultaneously and thus keep the distance 'D' between them constant, and he thinks that the rope should not break. So, does the rope break or not? The answer is that the rope does break. The explanation from the observer A's frame of reference is rather simple as we still have not considered the effects of length contraction. Due to length contraction, after the spaceships achieve a velocity v, every object contracts in length, including the rope. Therefore, although the distance between the ships remains the same, the rope itself shortens in length and hence breaks apart.

The Twin Paradox

The Twin paradox is probably the most famous in the Special Theory and sometimes even referred to as the mother of all paradoxes. If you remember, in previous chapters, we discussed a thought experiment where Mahesh ages 100 years while Geeta ages only one year when she goes on a trip in outer space. That situation was an example of the twin paradox. There, though, we could not convincingly explain what was happening in reality from Geeta's frame of reference, and that is exactly what we will do here.

Promoting space tourism, Geeta decides to go off and visit a star which is a distance of three light-years from Earth at a speed of 0.6*C (a light-year is the distance traveled by light in one year, it is a unit of length). Meanwhile, Mahesh stays back as he waits for Geeta to return. The Lorentz factor for Geeta's speed is 1.25.

According to Mahesh, Geeta takes ten years to cover her trip, five years up and down each. Mahesh also observes that Geeta's clocks run slower by a Lorentz factor 0f 1.25 because of the time dilation effect. Therefore, Mahesh agrees that while he ages ten years during Geeta's trip, Geeta should age only eight years on her return because time runs slower for her by a factor of 1.25. The paradox arises when we consider the same question from Geeta's perspective. According to Geeta, in her frame of reference, she observes that it is she who is at rest, but it is the entire universe that is moving backward with Mahesh and the star towards which she is headed. Now, in Geeta's frame of reference, because of length contraction effect, the entire universe contracts by a factor of 1.25 and therefore, the distance she needs to travel to the star system is also reduced by a factor of 1.25 and instead of 10 years she takes only 8 years to cover the entire trip. Furthermore, considering the time dilation effect from Geeta's frame of reference, time should run slower for Mahesh than for her, and she calculates that Mahesh should age only 6.4 years rather than eight years. Nevertheless, when she returns, she sees that Mahesh has aged ten years. Alice is left dumbfounded. What is happening? Where do the extra 3.6 years come from?

Now, let us consider this situation from only Geeta's frame of reference and break it down into two parts: outbound (when Geeta travels to the star) and inbound (when she returns from the star). During outbound, Geeta observes Mahesh moving backward at a speed of 0.6*C and according to her, $\Delta t(\text{Geeta}) = \dfrac{4}{1.25} = 3.2$, it takes the star 3.2 years to reach her. Similarly, during inbound, Geeta calculates that it would take Earth 3.2 years to reach her after she turns around, which sums up the total travel time to 6.4 years. Still, where do the extra 3.6 years come from?

To answer the question, it is important to note that after reaching the star, Geeta has to turn around to go back, or as she observes it, the star needs to turn around and recede away from her. Here, we need to remember that for Geeta to turn around, she has to first decelerate to stop and then accelerate in the opposite direction until she reaches the speed of 0.6*C. This is the tricky part. When Geeta decelerates and accelerates, she no more remains in an inertial frame (due to the introduction of acceleration and forces), and therefore, the concepts of Special Theory are no longer valid. Instead, in non-inertial frames, it is the General Theory of Relativity which can be applied. Now, according to the laws and math of the General Theory, Geeta should take 1.8 years to decelerate from her current speed to zero and another 1.8 years to accelerate from zero to a speed of 0.6*C in the opposite direction, which makes it a total of 3.6 years. Putting all the numbers together, Geeta takes 3.2 years to reach the star, 3.6 years to turn around, and another 3.2 years to reach back to Earth, which makes it a total of 10 years. This analysis explains the situation from both Mahesh and Geeta's frames of references.

Therefore, to understand the twin paradox, we must consider the factors introduced by the General theory, where we present the argument that Geeta is in a non-inertial frame that makes time run much faster for her, and she eventually takes 3.6 years just to turn around. We can also understand the twin paradox through the Special Theory of Relativity by using Lorentz transformations on spacetime graphs (a part of the theory we have not discussed to avoid complex mathematics in the book). The argument, in that case, is that during outbound, Geeta lives in a single frame of reference, but during inbound, when Geeta changes her direction, she inadvertently changes her entire frame of reference as well which is described by Minkowski spacetime and Lorentz transformations. Primarily, if we analyze this paradox solely using the Special Theory, we still find out that the time taken by Geeta to change her frame of reference is precisely 3.6 years, which can be explained using the concepts of General Relativity.

The twin paradox is a classic example of a paradox that employs both the sections of the theory of relativity (special as well as the general theory). A classic example that uses only the Special theory of relativity is that of the Pole in the Barn Paradox.

The Pole in the Barn Paradox

As the name suggests, this paradox revolves around a pole inside a barn. Here, Geeta is a swift pole vaulter, runs at 0.6 times the speed of light! Mahesh is a good old farmer who owns a barn with two gates, one in the front other at the back, and there are two atomic clocks at the respective doors. The length of the barn is eight meters, and the length of Geeta's pole is ten meters in their own frames of reference at rest. The lengths are hence proper lengths. The Lorentz factor for Geeta's speed is 1.25.

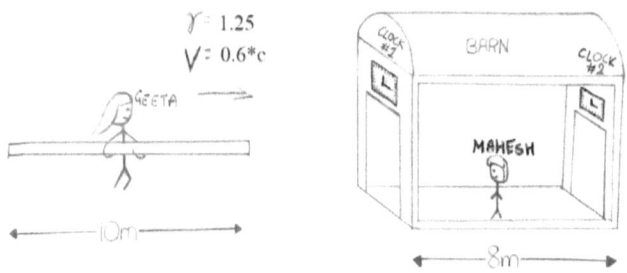

Figure 5.2 Representative image for the initial situation in the Pole in the Barn Paradox

In Mahesh's frame of reference, Geeta is moving at the speed of 0.6*C, and because of length contraction, he sees her pole contract to eight meters, $\Delta L(\text{Mahesh}) = \dfrac{10}{1.25} = 8$.

After observing this, Mahesh says, "Geeta, your pole is as long as my Barn. So, when you come running in through the front gate and completely enter inside, I will close the front door and open the back door for you to exit" (Mahesh is impeccable in his speed and we assume he can capably do that).

In Geeta's frame of reference, it is Mahesh who is moving backward at the speed of 0.6*C and she sees his barn contract in length to 6.4 meters according to the equation $\Delta L \left(\text{Geeta} \right) = \dfrac{8}{1.25} = 6.4$. By calculating the numbers, Geeta replies to Mahesh, "Are you crazy?! Your barn is just 6.4 meters and my pole ten meters long. There is no way my pole could completely fit in your barn." In saying so, Geeta is absolutely right. Although, in order to conduct the experiment, Mahesh is somehow able to convince Geeta to trust him and run through his barn. Hence, Geeta begins her run. She enters the front door, runs through the barn towards the back gate at speed 0.6*C and BOOM!

Well, that does not happen. Geeta does not collide into the gate as Mahesh successfully closes the front door, and in his frame of reference, the pole completely fits inside the barn. Simultaneously though, he also opens the back door just in time for Geeta to move out before banging into the gate. After completing her run safely, Geeta is dumbfounded. Although she is partly happy to be alive, she cannot understand how her pole fit inside Mahesh's barn when it clearly was much longer than the length of the entire barn.

Paradoxes to Ponder

Why did it happen?

Let us analyze the events one by one in separate frames of reference.

In Mahesh's frame of reference, as we have seen, things turn out perfectly fine. As soon as Geeta's pole entered the barn, he took a photograph of both the clocks (kept at the front and the back door of the barn), and as the pole reached the back door, he again took the photograph of both the clocks. According to Mahesh, Geeta runs at a speed of 0.6*C while covering eight meters inside his barn. So, the time taken by her to do that is: time = (8)/(0.6*C) = 44.4 nanoseconds. Now, as Mahesh compares the two photos of his clocks, he sees that in the first photo, both clocks show zero (because they were synchronized and set to zero) and in the second photo both clocks show 44.4 nanoseconds as in his frame of reference, they still remain synchronized.

In Geeta's frame of reference, after entering the barn at time zero, the back door takes $\frac{6.4}{0.6*C} = 35.6$ nanoseconds to reach her pole (length of barn in her frame of reference divided by the speed of barn moving backward for her). Also, for Geeta, time runs slower for Mahesh because of time dilation. Hence, when the back door reaches her, Mahesh's clock in Geeta's frame of reference reads

$$\Delta t(\text{Mahesh}) = = 1/\gamma *$$
(γ is Greek symbol gamma) nanoseconds.
$$\Delta t(\text{Geeta}) = \frac{35.6}{1.25} = 28.4$$

Another parameter to consider here is that of the leading clocks lag. For Geeta the barn is moving in the backward direction, so she observes that the clock near the back door runs faster than the clock near the front door and difference of time is given by $T = \dfrac{8*0.6*c}{c^2} = 16$ nanoseconds. Considering the leading clocks lag factor and the time dilation effect, we find that in Geeta's frame of reference, she sees Mahesh's clock read 44.4 nanoseconds only when the back door reaches her.

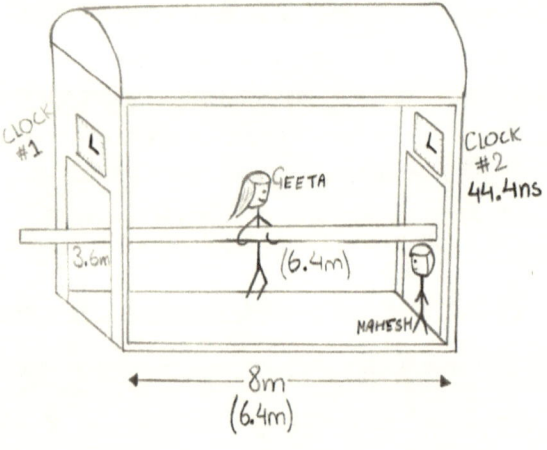

Figure 5.3 The situation in Geeta's frame of reference when her pole just touches the back door of Mahesh's barn.

At this point, the clock at the front door still reads 28.4 ns and 3.6 meters of her pole sticks outside of the barn in her frame of reference (the barn is 6.4 meters and her pole ten meters long). Now, because she has reached the back door and Mahesh has opened it just in time, the barn takes another 16 ns to reach the end of Geeta's pole. During that time, the clock at the front gate of the barn moves another 16ns to read 44.4 ns, and at that time, Geeta's pole would just completely enter inside the front door.

In this situation, we realize that the numbers fit perfectly well, and in Mahesh's frame of reference, Geeta's pole fits perfectly inside his barn. Though in Geeta's frame of reference, she never observes her entire pole entirely inside the barn at a single point of time, which can be explained by the relativity of simultaneity. Only because of the leading clocks lag factor, Alice can agree with Mahesh.

Other Paradoxes

There are many other paradoxes in the Special Theory of Relativity. One of them is the bug-rivet paradox, which is equally enthralling and seemingly even more hideous than the paradoxes we have studied. In this paradox, a bug lies inside a hole in a wall that is 10 centimeters deep. A rivet is used to kill this bug, and the length of the rivet is 8 centimeters. Now from a classical point of view, there is no way the rivet can kill the bug who takes a deep sigh of relief, but we know that we have superhuman capabilities. We do not strike at measly speeds of a couple of meters per second; instead, we strike at a speed of 223, 452, 105 meters per second for which the Lorentz factor is exactly 1.5.

In this situation, from the rivet's frame of reference, the hole in the wall contracts 1.5 times (gamma factor) to 6.6 centimeters, and quite clearly, you ask the bug to present its last wish as it is going to die soon. Our smart professor bug, though, refuses to die. It argues that because of length contraction effects, the rivet should contract in length by a magnitude of 1.5, and clearly, its new length of 5.3 centimeters is way shorter than ten centimeters; therefore, there is no reason for the bug to be afraid. You, who is smarter than both the bug and the rivet, are required to decide the fate of the bug. You are asked the million-dollar question – will the bug die or not? This is a tricky paradox and asserts that there is a contradiction in Einstein's theory. If you already have not realized how to solve this, I bring your attention to the similarity in this situation with that which we dealt with in the pole in the barn paradox. Figure that similarity out and you will be knocking on the door of the solution. An additional task which I suggest you should do is that if you tweak the speed with which the rivet strikes the bug a little bit here and there, you will notice that the answer to whether the rivet will kill the bug or not will consequently also change.

Apart from this, those who have heard of the detonator paradox, it is essential to note that that paradox is exactly same as the Bug-Rivet paradox in all aspects, with the only difference being that the bug is replaced with a detonator switch and the question becomes – will the rivet blow the detonator or not?

For the physics nerd: Apart from these paradoxes, the Ehrenfest Rotating Disk paradox is another deceitful paradox where we analyze a situation in which a circular disk rotates at a certain speed 'v.' Now, from a neutral observers point of view, the circumference of the disc (which is in motion) should contract because of length contraction and consequently the disc itself should contract to a smaller surface area as its circumference would decrease. But here is the catch, the radius of the disc is a perpendicular line segment to the direction of the motion of the disc, which is tangential with respect to the center. In that respect, we know that the radius does not undergo any contraction due to its invariance (discussed in the next chapter). So, here we are with a disc that logically should contract in size, and logically, its radius should also remain intact, but illogically, it still does not break apart due to the stress which should logically congregate across the disc. How can anyone explain this mess?

The solution to this mess comes from the fact that the Special Theory of Relativity only deals with ideal situations which do not involve any acceleration or presence of a gravitational field. In this situation, we have a circular disc rotating. It is essential to remember that the disc rotates because of the presence of a centripetal force which always acts towards the center of the disc and a centrifugal force that acts away from the center of the disc, thus dynamically balancing the forces. The introduction of forces, though, also introduces acceleration, which in turn creates a gravitational field, which brings us out of the mathematical and physical scope of special relativity into that of general relativity (discussed in the chapter '*What is so general about relativity*'). Once we step into general relativity in this paradox, we realize that the paradox can be answered using non-flat spacetime geometry and metrics such as the Schwarzschild metric. By doing that, all the effects such as time dilation and length contraction arising from the Special Theory will be countered by those which will be produced by applying the

General relativity. Hence, the effects produced from both theories would separately cancel out each other to create a paradoxical-less and logically consistent analysis.

Broadly concluding our study on paradoxes, I believe we mutually agree that Einstein's theory works just fine, and it is only the misinterpretation of concepts that gives rise to paradoxes. Still, after reading all of these paradoxes, you may come up with your own situations that might seem to defy the theory of relativity. In those scenarios, I motivate you to patiently go through each rationale in the case at hand. I assert that in the end, it will be Einstein who will triumph because of the geometry of Minkowski space-time (mathematics behind special relativity) that prevents any logical contradiction within the theory itself. If in case that does not happen, well, then it would be hard to beat you for the next Nobel Prize in physics.

CHAPTER 06

Special Theory in Real Life

"I seem to have been only like a boy playing on the seashore, and diverting myself in now and then finding a smoother pebble or a prettier shell than ordinary, whilst the great ocean of truth lay all undiscovered before me."

– Isaac Newton

"The whole of science is nothing more than a refinement of everyday thinking."

– Albert Einstein

"We live in a Newtonian world of Einsteinian physics ruled by Frankenstein logic."

– David Russel

"Research is what I'm doing when I don't know what I'm doing."

– Wernher von Braun

The theoretical aspects of the Special Theory of Relativity are quite convincing, which we saw by conducting numerous thought experiments, understanding various concepts, and analyzing different paradoxes. In physics, though, nothing is taken for granted, and no matter how much convincing a theory may seem on paper, it is of utmost importance that the theory must be robustly supported by experimental proof and experience for its firm establishment. Hence, we need to study a real-life situation where the Special Theory of Relativity can be applied to ensure that it practically does explain reality. A classic example, which confirms the relativistic perspective of the Special Theory, is that of the Muon, which presents a neat application of the theory in action in nature itself.

When cosmic rays from across the cosmos strike the lower atmosphere, they interact with the composition of gases and spray into various minute radioactive particles. One such particle is the Muon. A Muon is around 200 times heavier than an electron and has an average lifetime of about 2.2 microseconds before dissociating away (we know this through experiments conducted inside particle accelerators). These Muons are created approximately ten kilometers above the surface of Earth and travel at the speed of 0.9998 times the speed of light. By doing the math, one can quickly figure out that they should classically travel around 660 meters at the most before dissociating or dying away. This, though, is contradictory to what happens in reality because these Muons are observed at the surface of Earth and end up traveling around 10,000 meters against the predicted 660 meters they ideally should. So, what is happening here? The solution, as you may have guessed, lies in the Special Theory of Relativity.

Let us consider what is happening here. We become the observers by observing this phenomenon from a rest frame. For us, the Muons are moving at a speed of 0.9998*C, and therefore, time runs slower for them {time dilation}. As it turns out, the Lorentz factor for the speed of the Muon (0.998C) is around 15. Hence, the Muons live fifteen times longer than their actual lifetime. So, instead of traveling 660 meters, Muons end up traveling approximately 10,000 meters in our frame of reference, but what happens in the Muon's frame of reference? According to the Muon, it lives for only 2.2 microseconds but still ends up where it clearly should not. The thing is that for the Muon, it would see the Earth's atmosphere traveling at 0.9998C, and because moving lengths contract, it would see the atmosphere of Earth contract 15 times its original length. Therefore, 10,000 meters contract into around 660 meters (a rough measure), and the Muon can travel those 660 meters in 2.2 microseconds to end up at the Earth's surface.

Conclusively, from a general observer's frame of reference on Earth, it is the time dilation effect that makes the Muon's journey permissible, and from the Muon's perspective, it is the length contraction effect which helps it to travel 10,000 meters in place of 660 meters. In both cases, though, the theoretical analysis matches up with the reality as we have to have it, and the Special Theory eventually triumphs.

Only for the sake of simplicity, we have discussed one example out of many situations where the Special Theory can be neatly applied in real-life. Another example which we will discuss later is that of the Global Positioning System or the GPS as it is better known.

What Is Not Variant?

Until now, with Einstein's help, we have learned that time and length are not absolute but relative measurements, and even though each of us carries our own clocks that are equally precise and meticulously synchronized when we move relative to one another, these clocks disagree to agree. Similarly, we all carry our own yardsticks, and each of them is precisely of the same length, but when we move relative to each other, they just do not agree. Another third popular idea of relativity is that of the relativity of mass. It is generally said that mass increases as we go faster. Furthermore, this statement is also used to explain why we can't go faster than the speed of light and the reason given is that because larger the mass of the object, larger is the energy required to accelerate it, and near the speed of light the mass of an object will approach infinite and hence an endless amount of energy will be required to speed it. Thus, one may conclusively announce that due to the relativeness of mass, it is not possible to achieve speeds faster than the speed of light.

The understanding that mass is relative and increases with motion is contextually flawed, and I assure you that there is no such increase in mass whatsoever. It remains constant throughout, but what increases is relativistic mass.

When we talk about relativistic mass, it is vital to know the origins of the term. Momentum is a physical quantity mathematically defined as $p = m * v$; where p is the momentum, 'm' is the mass, and 'v' is the velocity of an object. In essence, momentum provides us the measure of the impetus gained by an object. The equation for momentum was given by Newton, but the actual equation, given by Einstein, is $p = \gamma * m * v$, where gamma is the Lorentz factor. In typical day-to-day situations, the value of γ remains exceptionally close to one, and Einstein's equations become equivalent to that of Newton's. However, close to the speed of light, when gamma becomes significantly large, Einstein's equation gives us accurate answers. In that scenario, to make the equation look simpler than it actually is, scientists often combine 'γ' and 'm' into a single term '$\gamma * m$,' which is named as *relativistic mass*. This integration makes a physicist's work straightforward. Hence, whenever a physicist talks about or introduces the term relativistic mass, that person is actually referring to the combined term of the 'gamma factor' times the 'mass' of the object. It is relativistic mass that increases with speed, but the term is misguided and does not refer to proper mass itself. '*Proper Mass*' is a term often used in the scientific community and refers to the original mass of the object that remains constant in all frames of reference.

For those who have read the book '*A Brief History of Time,*' in the second chapter Hawking uses the Energy-mass equivalence relation, $E = mc^2$, to explain that faster the speed of an object is, higher is its mass, and hence greater is the energy required to speed it up. It is vital to note here that although he mentions mass, Hawking still refers to the notion of the relativistic mass because the actual equation is $E = \gamma * mc^2$. Just for simplification, '$\gamma * m$' is called relativistic mass, and it is this quantity which rises with the rise in speed but not mass itself. Conclusively, we can say that mass is not a variant quantity.

After reading about length contraction, a question naturally arises in the mind of the reader – if length is relative, then is it that height and width are also relative measurements and depend upon the frame of reference. The answer is no. It is only the length of an object that is relative, but *height and width are not affected by relativistic effects* or the frame of reference of the observer.

To understand why that happens, imagine a train car moving on rail tracks while you stand and observe it moving from the ground. For the sake of understanding, also assume that width is a relative quantity and must shrink for moving objects. After considering that, as an observer on the ground, you should clearly observe the train wheels shrink and fall inside the rail tracks, but that does not happen. Maybe our assumption was wrong and we say that although width is a relative quantity, it must expand for moving objects. Through this conjecture also though, we do not see the train wheels expand and break away from the tracks, which again dismisses the possibility of relativeness of width.

Similarly, let us consider a situation where a train is moving inside a tunnel. We assume that height is a relative quantity and increases with the increase in speed. From an observer's frame of reference standing on the ground, you should observe that the train's height should increase and it should crash into the head of the tunnel, but that clearly never happens. Again, maybe our assumption was wrong and we say that although height is a relative quantity, it must decrease with the increase in speed. In that case, consider yourself to be sitting inside the train. Now according to your frame of reference, you and the train are at rest, but it is the tunnel which is moving backward, and hence, as per our assumption, its height should decrease. That too, though, never happens because the tunnel never crashes into the roof of the train. This analysis dismisses the possibility of relativeness of height.

One may argue that just like the observable effects of time and length, the relative effects of height and width should also become significant only at near-light speeds. In considering this, you are right, but it has been mathematically proved that width and height are not relativistic quantities like time and length, and there is conclusive experimental evidence to support that claim. These quantities which are not relative are called 'invariant' and are helpful to understand 'invariant intervals,' in-depth explanation of this term is beyond the scope of this book.

Faster than Light

If you remember, we previously discussed that nothing could go beyond the speed of light. The explanation for that statement comes through the Lorentz factor given by $\gamma = \dfrac{1}{\sqrt{1 - \dfrac{v^2}{c^2}}}$. In this expression, if we put the value of v as c, it turns out that the term inside the square root not only becomes zero, γ but also becomes infinite. Now in mathematics, infinite is not a number that exists on paper, but it is only an idea. Thus, when equations start giving out answers like infinite, mathematics is considered to break down, and therefore numbers like 1/0 are prohibited by the rules of mathematics because their logical value comes out to be infinite, an unspecific inexistent quantity. Similarly, in this situation, we cannot practically explain what happens at the speed of light because equations give infinite as an answer, which is out of the scope of mathematics. Hence, two solutions are proposed. One, Einstein's theory does not work. Second, there is an ultimate speed limit. Now, because Einstein's theory is well proven and experimentally robust, we conclude that the first option is invalid and the second option is the solution. Therefore, there must be an absolute speed limit, which is simply the speed of light. Technically speaking, no object can travel at the exact speed of light leave alone faster than it.

After making this conclusion, though, there were critiques, as there always are, who claimed that although we cannot reach the speed of light, there must be something which moves faster than it. It was, hence, proposed that maybe there is a spectrum beyond the speed of light, probably another world or reality which we just cannot see, and these hypothetical particles, christened Tachyons, travel faster than the speed of light in that other reality. A significant amount of theoretical work was conducted in this field to figure out if something like the Tachyons could exist, but the theory ran into an insurmountable amount of problems and contradictions. As of the time of writing of this book, Tachyons remain to be hypothetical particles and are of noteworthy interest only to science fiction authors as there is no experimental proof and only little theoretical proof which suggests but not entirely supports their existence.

Freezing of Time

To make things fun, we assume that it is actually achievable to attain the speed of light if not go beyond it. To work out how that situation may pan out, let us conduct another thought experiment. We make use of a light clock given in figure 6.1, with mirrors A and B parallel to each other, and set the entire clock into motion at the speed of light.

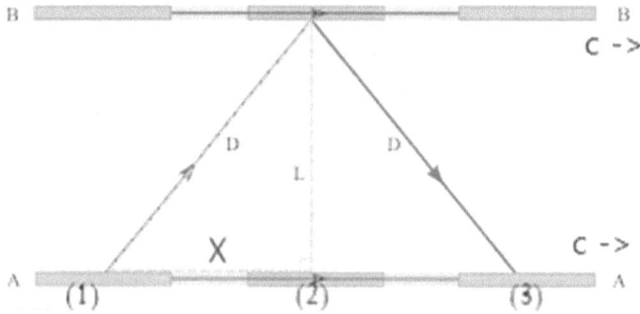

Figure 6.1 A light clock moving towards right at the speed of light 'c'.

At time t=0, the light ray leaves from mirror A in position one to reach mirror B in position 2. From a stationary observer's frame of reference, the mirror B needs to travel a distance X while the light beam needs to travel a distance D to meet at position 2. Both the mirror and the ray are traveling at the speed of light (c), and quite evidently, one can see that D is greater than X. This means that when the light ray reaches position 2, the mirror would have already gone further away. Therefore, for the light beam to actually meet the mirror, it would have to go further on and by careful analysis, you can figure out that the light beam would never reach the mirror B because both are traveling at speed c and both of them will travel farther and farther but never actually meet each other. Therefore, the observer at rest never sees the light beam reach mirror B, and to that same observer, time seems to have frozen entirely inside the light clock.

So, if Geeta is asked to board her spaceship and travel at speed c, while Mahesh observes her from his frame of reference, Mahesh will see that *time freezes for Geeta*. From Geeta's perspective, though, time runs just fine. She sees that it is not her but the entire universe which is moving back at the speed of light. This is a situation similar to the twin paradox. We can resolve it by saying that because Geeta needs to accelerate to reach the speed of light and then decelerate to stop, both the events themselves would take an infinite amount of time to complete. So, even after traveling for perhaps only one nanosecond in her frame of reference, Geeta would reach the end of everything and arrive probably beyond the end of time itself. She might age billions, trillions, zillions of years; who knows how old she may be or what she may find at the end of time? Unfortunately for Geeta, though, Einstein prohibits her from doing anything like that by banning the ability of anything to travel at light speed. This ban makes Geeta's adventure one filled with fantasy but with much scope in the genre of science fiction.

After vigorous assertion of the applications of Einstein Special Theory of Relativity, as provided till now, people still come up with ideas and situations which seemingly oppose the theory. One such scenario is that of velocity addition. Suppose that Geeta is moving at a speed of 200,000 km/s for observer Mahesh who is at rest. Now, she somehow can throw a ball at a speed of 150,000 km/s, and because of velocity addition, Mahesh should observe that the ball is traveling faster than the speed of light, yay! However, that does not happen. The basic definition of velocity is the distance covered divided by time. In this case, because of length contraction, time dilation, and leading clocks lag effect, the velocity of the ball will still never go faster than the speed of light, according to Mahesh in his frame of reference. Lorentz transformations and other equations are employed to figure out the actual velocity of the ball, which is out of the scope of this book.

Equation of the Century

What does anyone do when they derive the equation of the century? Einstein did not think of the several t-shirts it would be printed on as he silently sat on his living room couch with a cup of coffee lost in the implications of what he had just derived.

$E = mc^2$

('E' energy; 'm' mass; 'c' speed of light)

Undoubtedly the most famous equation after Newton's law of gravity, Einstein's energy equation, or the mass-energy equivalence principle as it is better known as, is familiar to almost each one of us today, but lesser are the people who realize its implications. Einstein came up with the equation in November 1905 after he published his Special Theory of Relativity, but it was more of like 'Oh, and I forgot to tell you this' kind of addition. He said that the equation arises from the consequences of symmetries of space and time itself, which he, unfortunately, realized only after publishing the Special Theory.

At first sight, the equation looks overwhelmingly simple. In fact, it seems so simple that I wish my high school math equations were more like these than the complex differentials, integrals, summations, and whatnot. However, when given profound attention, one finds that the equation's daunting meaning is nothing less than revolutionary. The equation tells us that mass and energy are different ways of measuring and understanding the same thing. Think it this way; just as ice is frozen water which spills everywhere when melted, mass is like *'frozen energy'* waiting to break the shackles and spread out, just not by melting though. This profound meaning of the equation provides us with an entirely new dimension to explore as it tells us that mass is actually only another form of energy as they are, in fact, interchangeable quantities. Mass can be broken apart into different forms of energy, whereas it can also be produced out of pure energy. So, if an object is heated by say 10 degrees, that object actually gains mass which can be calculated by $m = E/c^2$, which, in this case, will turn out to be negligibly small due to the c^2 factor, but still, there is an increase in mass.

When Einstein devised this equation, his purpose was to understand the nature of reality and not create atom bombs 40 years down the line. That is right, although the involvement of the equation in the Manhattan project was very little, this tiny three-letter equation still forms the basis of how atom bombs get their devastating energy. When neutrons are fired at plutonium or uranium, the heavy atoms split apart into lighter ones. Careful analysis shows that in this incredibly fast transition, mass is not conserved, and the small amount of mass which is lost is converted into energy, which can be calculated using Einstein's equation. Such immense is the energy produced by this minor mass defect that one atom of uranium can produce energy equivalent to more than ten million times that produced by heating a single atom of coal.

Putting the significance of Einstein's equation into simpler words, an average 70 kilograms heavy human body carries the energy equivalent to 2500 Hydrogen bombs, which is enough to power more than a billion homes for over three years continuously. In general, the energy in an average human body is over 80,000 times more than the bomb that killed hundreds of thousands of people in a matter of few seconds when it flattened out Nagasaki in Japan 1945. So, the next time you say to yourself, *'I do not have any energy left,'* you should remember exactly how much energy you have got in the tank, however, if we knew how to use the immense amount of energy we all possess, who knows how the world might have looked like today.

Light Cone

When light is released from an object, it travels in all directions and spreads out like an ever-growing sphere. An analogy to relate this with is that of ripples, just like a ripple of light that is also a wave and spreads out in a circle from the point of origin. Mathematically speaking, in one second light from a particular object will have traveled 300,000 km across in each direction; in two seconds, the same wave would have traveled 600,000 km in all directions, and so on, it will keep increasing.

The graph in figure 6.2 (b) is a spacetime graph, also known as a light cone. The path of light rays (which form a cone-like shape) mark the boundaries until where light can travel from the center, beyond these rays, light cannot travel; these lines are also called world lines. If an event occurs at position B, its effects can be observed by the observer at the junction of the axes at point O. All such events which lie inside the boundaries can affect the observer at O, and this interval is known as a timelike interval. Hence, events occurring at point E and B can affect the observer. The difference between them is that B occurs in the future, whereas E has already occurred in the past for the observer. All the events which occur on the world lines are called lightlike intervals and can influence the observer O, only if the information of the event can travel at the speed of light to just reach the observer, for example, event C. If an event occurs at any other position than we just discussed, it is called a spacelike interval, and sadly, it can never influence the observer at O. For example, event A and event D occur in spacelike intervals and cannot influence the observer at O, because the information cannot travel faster than the speed of light. The graph in figure 6.2 (a) depicts the situation in 3-dimensions. Hyperspace is a term used to specify a plane in a particular frame of reference. Here it refers to the present plane of the observer.

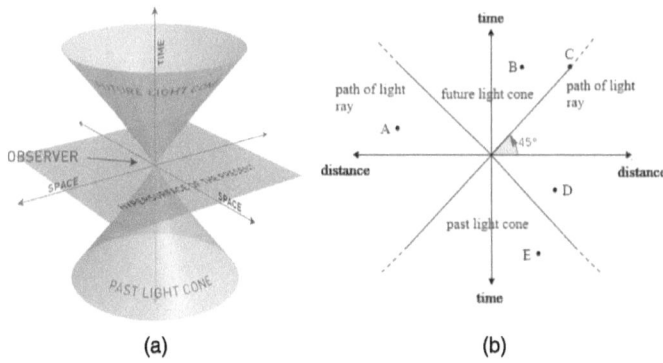

Figure 6.2 (a) A 3-dimensional light cone. (b) A 2-dimensional light cone.

Each year, literally thousands of claims are made by students and scientists across the globe who claim to have proved one of Einstein's postulates or even theories wrong. Therefore, it is not that people are not studying Einstein's work, they are very much investigating the idea of Tachyons and other theories where they imagine, analyze, and come up with unprecedented situations that seemingly violate Einstein's theories. Though, until today, further analysis has shown that due to minor errors or the presence of some mitigating factors, Einstein's theories eventually hold true. However, this also does not mean that the theory is absolute or marks the end of physics. There may be some theory or experiment performed in the future, which could potentially overturn our understanding of the universe once again by giving a different form and explanation to our reality.

CHAPTER 07

What Is so General about Relativity

"Gravitation is not responsible for people falling in love."

– Albert Einstein

"You cannot teach a man anything; you can only help him discover it in himself."

– Galileo Galilei

"One could perhaps describe the situation by saying that God is a mathematician of a very high order, and he used very advanced mathematics in constructing the universe."

– Paul Dirac

"The history of science shows that theories are perishable. With every new truth that is revealed, we get a better understanding of Nature, and our conceptions and views are modified."

– Nikola Tesla

After publishing his Special Theory of relativity, Einstein realized that the situations in which his theory could be applied to were those where objects did not accelerate or were bereft of any influence from an external force. This was a crucial limitation in his theory because, according to Newton's gravity, if between two objects, one is moved, the other would instantaneously experience the change in the gravitational force. This was a well-established and commonly observed phenomenon but also implied that gravitational effects should travel at an infinite velocity. Einstein recognized this chink in the armor of the Special Theory. Hence, to make things right, he started working out parameters to build another more universal or a more general theory, which would take into account gravity while still being consistent with the current principles of his Special Theory and physics in general.

After reflecting on the nature of gravity itself, Einstein conceived that the behavior of gravity was analogous to that of a magnetic field. When a magnet attracts an iron piece, we do not regard the action as a direct consequence of some interaction between the magnet and the iron piece instead; we say that the magnet induces a magnetic field in the empty space around it which in turn makes the iron piece attract towards the magnet. Similarly, he said that when a stone falls on Earth from a particular height, the action of the gravitational force of Earth on the stone takes place indirectly. Earth produces a gravitational field in its surroundings, and it is this field that is solely responsible for the stone to fall. From this insight, Einstein determined that the law which governs the properties of gravity in space is due to the gravitational field rather than a direct interaction of gravitational forces. Moreover, gravitational fields exhibited a remarkable feature in contrast to other fields such as electric and magnetic fields, because it does not depend on the material or physical state of the body it acts on. This understanding prompted Einstein to think of the gravitational field as a consequence of the integration of space and time, which, according to him, were fundamental to the very nature of the fabric of the cosmos.

This conception of the field was not new for the physics community. Even today, we encounter several types of fields, including nuclear fields, Quantum fields, Higgs fields, among others, which are integral to the formulation of modern physical laws. To understand this concept of a gravitational field, Einstein came up with numerous ingenious thought experiments when he simulated all the possible scenarios up in his head.

The 'General Theory of Relativity,' in essence, is a geometrical theory of gravity which accurately describes the workings of the cosmos at a macroscale. The theory is background independent and is based only on a set of sixteen equations known as the '*Einstein Field Equations.*' These field equations, provided by Einstein in his theory, are complicated to solve and have a fixed number of solutions. Many of these solutions, such as '*Gödel Universe*' (talks about the possibility of time travel), have not been physically exploited due to technological limitations but are theoretically valid. In this chapter, we will go through only some of the concepts, ideas, and results of the General Theory of Relativity as an in-depth analysis is way out of the scope of the book.

The Happiest Thought

During later years of his life, while reminiscing about his thought experiments, Einstein wrote in German, *"I was sitting in a chair, in the patent office in Bern, Switzerland when all of a sudden a thought occurred to me. If a person falls freely, he will not feel his own weight. This revelation startled me."* This thought, as he later wrote, lead him towards the *"happiest thought"* of his life.

Imagine: A person is standing in an entirely closed elevator somewhere out in deep space. Due to lack of gravity, the person will not be attracted to the ground of the elevator, and he would not even feel his own weight. At this point, if the person drops a ball, the ball would float in the air and stay there because there is no gravity pull it down. Now, the elevator is attached to a crane (astonishingly transported to deep space), which pulls the elevator with some acceleration. During this upward acceleration, the person inside would suddenly feel heavier as he feels a momentary push downwards. He would feel as if gravity has suddenly made an appearance and was now pulling him towards the floor of the elevator. Moreover, this person inside the elevator would not be able to tell if he is on Earth or outer space because the elevator is walled, and the acceleration acts like gravity for him. Therefore, an acceleration upwards in the elevator is equivalent to the presence of gravity. This thought was Einstein's **happiest thought.**

However, why was the happiest thought so happy?

As stated before, after publishing the Special Theory, Einstein was mulling over how to categorize motion in the presence of gravity. The 'happiest thought' implied that the dynamics of a situation produced in an accelerated frame of reference would be similar to those produced in a gravitational field. This insight was extremely happy for Einstein because now he could evaluate the dynamics of an event in an accelerated frame of reference, which is doable and straightforward, and apply those same results in a gravitational field where they will be equally valid. This concept today is known as the *equivalence principle*, which essentially means that the effects of an accelerated frame of reference are also applicable in a gravitational field.

Eventually, this thought was the first of many similar analyzations and insights, which helped Einstein integrate his previously established principles with the concept of gravity and produce what we today call the General theory of relativity.

Gravitational Time Dilation

An elevator is placed deep in outer space. It has a combination of two light beams with transmitters and receivers, each attached at the top and bottom inside the elevator. The transmitter sends light pulses from one end of the elevator to the other at regular intervals. Two clocks, clock L and clock U, are also placed at the bottom and the top of the elevator, respectively, as shown in figure 7.1.

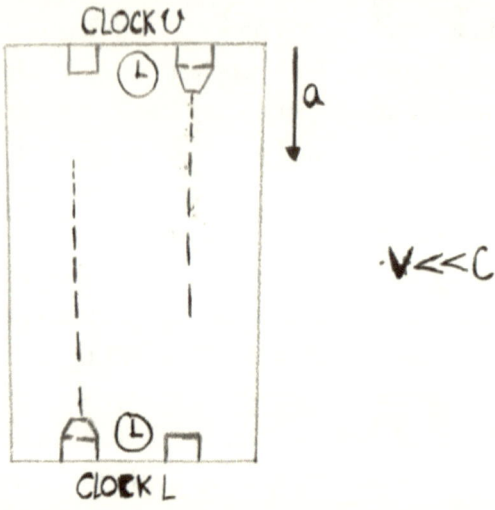

Figure 7.1 The elevator in outer space accelerating downwards with acceleration 'a'.

Initially, the elevator is at rest to an observer but starts accelerating downwards at an acceleration 'a' to achieve a speed 'v.' To nullify the effects of the Special Theory of Relativity, we assume that the speed v is much smaller than c. Considering the transmitter at the top, let us say it sends ten pulses per second to the receiver at the bottom. The receiver, though, is accelerating downwards and therefore receives eight pulses of light per second as the light pulses have to cover a longer distance to reach the transmitter, which is accelerating away. In the second pair of transmitter-receiver, the transmitter at the bottom shoots only ten pulses of light per second, but the receiver at the top receives twelve pulses of light per second. This happens because the top of the elevator is accelerating towards the light pulses, which then have to cover less distance to reach the transmitter.

Hence, to an observer at rest watching the experiment from outside, he would perceive that the clock L, the bottom one, is slower than the clock U at the top, because the receiver at the bottom receives lesser light pulses per second than the receiver at the top, whereas the light pulses from both the transmitters (top and bottom) are transmitted equally at the same speed. Through the equivalence principle (discussed before), this situation of the accelerating elevator can also be implied to an elevator which is at rest in the presence of a gravitational field. Now in a gravitational field, acceleration acts towards the source of the field. Therefore, in this case, because the elevator is accelerating downwards, the gravitational field also acts downwards, i.e., from the upper end of the elevator to the lower end. Hence, one can conclude that time runs slower at the lower end of the elevator compared to time running at the upper end of the elevator because gravitational force (which induces acceleration) is stronger at the bottom than at the top. A general statement that can be made from this conclusion is, *'time runs faster as one goes further away from a gravitational field, or closer you are to the source of a gravitational field, slower the time runs for you.'*

This gravitational time dilation effect remains unobserved in our everyday lives because the difference in gravitational force at ordinary altitudes on Earth is negligible. This effect, though, is exceptionally significant at higher elevations and becomes essential to be taken into account in the case of the Global Positioning System (GPS).

The GPS satellites fly approximately 20,000 kilometers above sea level, and as they are much farther away than us in Earth's gravitational field, time runs faster for them and leads by approximately 45 microseconds each day (effect of general theory). Moreover, the satellites above run much quicker than the speed with which Earth rotates around its axis, and hence, we also need to consider the time dilation effects from the Special Theory, which are independent of the impact of the general theory. On doing the math, it turns out that to an observer on Earth, time runs slower on the GPS satellites and lags by approximately 7 microseconds each day (effect of Special Theory). In totality, by combining the results of both the theories, we learn that time runs ahead on the GPS satellites in orbit by approximately 38 microseconds each day.

This time lag may seem to be small, but it is incredibly significant. After doing the math, it is found that if the effects of time dilation are not taken into account in the operating system, the satellites will start producing inaccuracies in determining positions at the rate of approximately ten kilometers each day. Simply put, if today the GPS correctly shows that a person is in his home, in the same position a week later, the GPS would show the same person in another city altogether. Hence, for the successful running of the GPS, it is essential to take the effects of time dilation into account, and by doing that, as of today, GPS produces a maximum inaccuracy of around five meters in detecting a specific location.

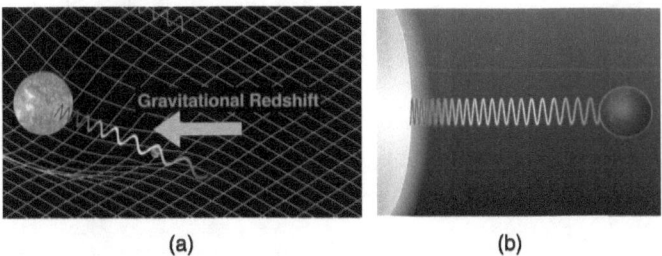

Figure 7.2 (a) Gravitational redshift of light rays traveling through depression in spacetime. (b) 2-dimensional depiction of gravitational redshift.

Gravitational time dilation also gives rise to gravitational redshift. This essentially means that as a wave travels away from a gravitational field, its wavelength redshifts to larger wavelengths as the gravitational pull acting on the wave decreases and the wave itself stretches to longer wavelengths (red light has the longest wavelength). Vis' a vis' if the wave is traveling towards the source of the gravitational field, its wavelength will blueshift towards smaller wavelengths as the effective gravitational force acting on the wave will increase (as shown in figure 7.2), and the wave would compress. This effect is crucial for scientists to evaluate the distance of cosmic objects. The more redshifted the light is from an object, the farther away the object lies, and the more blueshifted the light from an object is, the closer it lies to us.

Warps and Gravity

Figure 7.3 An artistic depiction of depressions in spacetime where Earth is revolving around the sun.

If you roll a ball across a smooth marble floor without any acceleration, it will travel in a straight line. If you roll the same ball in the same way on a sandy surface with little holes and bumps, the ball will not go in a straight line as it would roll in and out of that irregular surface in a curved path. Einstein imagined the fabric of the universe in the same way. He said that the universe lies in this spacetime (can be imagined like a fabric), which is extremely flat and smooth like the marble floor, but in the presence of matter and energy, the same spacetime floor behaves like a sandy surface with warps and curves. So, matter and energy are responsible for curving spacetime proportional to the amount of mass of the matter or the energy present. Similarly, the sun bends a region of spacetime around it, and Earth moves in that sort of a ditch around the sun, as shown in figure 7.3. This, according to Einstein, is precisely how gravity works.

On a flat surface, Earth would travel in a straight line, but on this curved surface, it follows the nearest path to a straight path, called a geodesic, which is the shortest path between any two points in a curved surface. For an analogy, an aircraft travels across the globe, not in a straight path, understandably, but in a geodesic path, which is the shortest path between two airports. In absolute terms, matter travels along a straight line in four-dimensions, which is the true nature of spacetime but seems curved for us because we are capable of observing spacetime in only three-dimensions. Analogously, if you see an airplane go over a hill, you would see the plane travel in a straight line in your three-dimensional view, but its shadow takes a curved path on the hill as the shadow moves in only two-dimensions.

By doing the math after predicting such nature of gravity, Einstein was able to explain the orbits of all the planets with the highest accuracy confirmed by the present-day data. In contrast, Newton's predictions were approximate but not as accurate as Einstein's. In conclusion, according to the general theory, the geometrical properties of space, which include warps and curves, are not independent quantities but are determined by matter and energy themselves. Therefore, we can understand the structure of the universe by mapping out quantifiable entities such as mass and energy, which may be present regularly or irregularly across the fabric of space and time itself.

Light in Spacetime

Imagine an elevator that accelerates from rest at an acceleration 'a.' A hole in the top of the elevator allows a pulse of light to come in and is observed by Geeta inside the elevator and Mahesh, who is outside. We divide the situation into three phases, as shown in figure 7.4 in Mahesh's frame of reference.

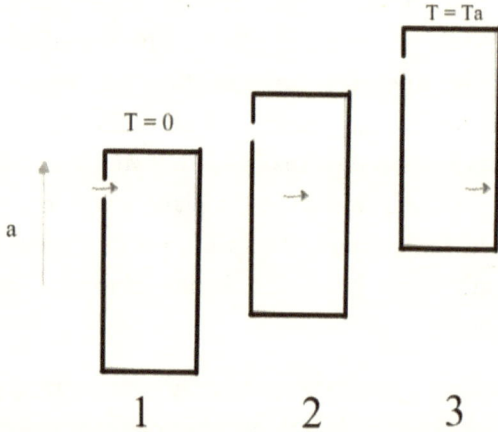

Figure 7.4 Light ray travelling through Mahesh's elevator from time T=0 to T=Ta as observed by Mahesh from position 1 to 3.

In the first position at time T=0, Mahesh sees the light enter the elevator from the top. In the second and third positions, the elevator accelerates upward at acceleration 'a' as the light passes in a straight line to hit the elevator at the bottom in the third position at time T=Ta. In the same situation from Geeta's perspective, she sees the light bend over a period of time from T=0 to T=Ta, as shown in figure 7.5.

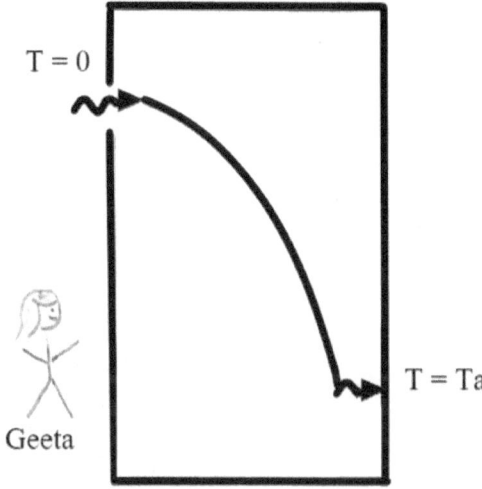

Figure 7.5 Light ray bending in Mahesh's elevator from time T=0 to T=Ta as seen by Geeta from her frame of reference

In conclusion, light bends in an accelerated frame, and through the equivalence principle, this conclusion is also valid in a gravitational field. Therefore, light bends in the presence of a gravitational field. This was a momentous outcome of Einstein's theory. He said that just how planets revolve around the sun due to the depression, it creates in the fabric of spacetime, light also bends and travels in a geodesic path wherever matter creates similar depressions like the sun. This phenomenon is also known as gravitational light bending.

This effect of light deflection, in turn, gives rise to the effect of gravitational time delay. This effect explains that because light is deflected in a gravitation field, it travels in a geodesic path (the shortest route to travel in four dimensions), but contrastingly, in the absence of a gravitational field the light signals would travel in a straight line which would always be shorter in absolute distance when compared with a geodesic path. So, as light has to travel farther in the presence of a gravitational field because of a geodesic path, it results in a time delay which is coined as gravitational time delay or Shapiro delay, after the person who postulated it.

Final Comments

Einstein published his theory on gravity with final finishing touches in 1915, and one of his predictions was that light bends under the influence of gravity. Though, to observe this effect, one required a strong gravitational field. A method was proposed to verify this. During the day, stars present behind the sun are understandably not visible to us, but if Einstein's theories are correct, their light must bend, and it would seem that they may lie just at the edge of the sun. This phenomenon is called gravitational lensing effect and is discussed in more detail in chapter ten, holes and waves in the dark cosmos; the idea is visually depicted in figure 7.6. In order to get data on gravitational lensing, scientists had to wait for a total solar eclipse to observe the light coming from those stars near the edge of the sun, which otherwise were dwarfed in front of the sun's glare.

In 1919, a total eclipse was supposed to occur off the coast of Africa and Brazil. During then, the World War had also stopped, so the British decided to send two teams to both the locations for observations. A similar attempt was made in 1918, but due to bad weather and clouds, sufficient data could not be collected. Though, as luck had it, Sir Arthur Eddington's expedition to the island of Principe near West Africa was able to collect sufficient data of the stars near the sun's edge before coming back to England to study the results. The work was extremely tedious and demanded high accuracy as the scale of the data was relatively very small; they had to take into consideration the temperature difference, any movement in the lens of the telescopes, or the tripods, among many other potential errors. In simple words, a single mistake could either destroy or establish Einstein's theory. After months of calculations, analysis, and comparisons, Sir Arthur finally claimed that Einstein's theory triumphed. In a meeting of the Royal Society of London, he claimed that Einstein's theory accurately predicted the amount of light which should bend around the sun's gravitational field as those predictions comfortably coincided with his experimental results with minimal errors. The errors which were produced were within acceptable limits and could be because of minor experimental limitations.

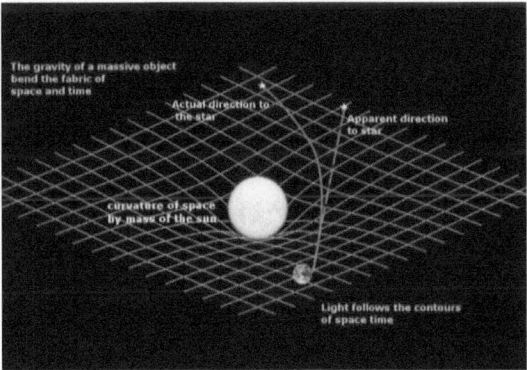

Figure 7.6 Bending of light from a distant star around the sun due to sun's gravitational influence.

This was a sensational event because the world was weary of a war where the British and the Germans were bitter enemies. However, here comes a British expedition confirming a German scientist's theory overturning the work of Britain's or arguably the world's best scientist till then. The media caught the wind of these results and highlighted Einstein's success over Newton around the globe, making Einstein a scientific icon and a sensation around the world. His fame, quite naturally, initiated debate amongst the general public and the scientific community as to why Einstein was not being awarded the Nobel Prize. He was regularly selected for the prize since 1910 but could never win it. The committee which presided over announcing the results included primarily of experimentalists who were skeptical about Einstein's theory, given that it was not robustly supported by experimental proof. Therefore, although Einstein was awarded the Nobel Prize in 1921, he was not awarded it for his work on the theory of relativity but instead for his work on the photoelectric effect, which he talks about in one of his papers published during 1905, the miracle year.

The committee stated that once they would receive more proof supporting Einstein's theory, they would award Einstein the Nobel Prize once again. This, unfortunately, never happened as the theory of relativity was firmly proved only in 1971 through the Hafele-Keating experiment. Hafele and Keating took four Cesium beam atomic clocks in a commercial airliner across the world, first eastwards then westwards, in a bid to observe time dilation effects as described by Einstein. After completing the experiment, they compared the results between their clocks and others that remained at the United States Naval Observatory. Hafele and Keating found that the clocks actually did disagree between them, and the difference in the readings of the time in the clocks was calculated to be consistent with the predictions laid down by Einstein's Special and the General Theory of Relativity combined.

Other predictions of Einstein's general relativity like the existence of Black Holes were first confirmed in 1971, and the presence of gravitational waves could only be established until 2016. It took almost a century to produce equipment sensitive enough to do so. The theory's other predictions, such as the existence of white holes and wormholes, among others, remain entities for science fiction because their presence has yet never been experimentally recorded or observed through even the most powerful telescopes produced till date. Einstein's legacy is one for the ages as humanity, even today, follows his footsteps to discover what the universe has to offer to us.

The Cosmos in a Platter

"I think nature's imagination is so much greater than man's; she's never going to let us relax."

– Richard Feynman

CHAPTER 08

From the Beginning

"Today we're still loaded down – and, to some extent, embarrassed – by ancient myths, but we respect them as part of the same impulse that has led to the modern, scientific kind of myth. But we now have the opportunity to discover, for the first time, the way the universe is in fact constructed as opposed to how we would wish it to be constructed."

– Carl Sagan

"The strength and weakness of physicists are that we believe in what we can measure. And if we can't measure it, then we say it probably doesn't exist. And that closes us off to an enormous amount of phenomena that we may not be able to measure because they only happened once. For example, the Big Bang."

– Michio Kaku

"The important thing is not to stop questioning. Curiosity has its own reason for existence. One cannot help but be in awe when he contemplates the mysteries of eternity, of life, of the marvelous structure of reality. It is enough if one tries merely to comprehend a little of this mystery each day."

– Albert Einstein

"Look up at the stars and not down at your feet. Try to make sense of what you see and wonder about what makes the universe exist. Be curious."

– Stephen Hawking

Birth of Cosmology

The earliest origins of the idea of cosmology date back to more than two millenniums ago. Aristarchus of Samos, a Greek philosopher, presented the first heliocentric model of the then-known universe wherein he placed the sun in the middle of the model with the earth and other planets revolving around it. But Aristarchus's model, on account of general opinions and daily observations, was not received well. Astronomers like Aristotle and Ptolemy, who worked towards celestial motion and philosophized the workings of the known universe, proposed the geocentric Ptolemaic system, which was robustly accepted. In this system, Ptolemy placed Earth at the center of the world while all other celestial objects like the sun, moon, and stars were made to revolve around it. This universe was thought to be finite with a defined boundary that marked the end of the stars, and the followers of the Ptolemaic system firmly believed that beyond the stars, the heavens revolved around Earth. This idea also explained that God always keeps a watch on earthlings, a concept which was later modified by the church, declaring that God is the creator of everything known.

This model of the universe predominated until the dawn of the 16th century when Nicolas Copernicus brought forward his heliocentric model of the universe with the sun being the center and the Earth revolving around it. Copernicus's revolutionary work sparked the Copernican revolution that was carried forward by Kepler, Galileo, Newton, and others. Kepler extended Copernicus's work by studying the planetary motion and introduced Kepler's laws, which essentially described that even though Earth revolves around the sun, it revolves not in a circular but an elliptical orbit with the sun placed at any one of its foci. Through the mathematical formulation of planetary orbits, Kepler was able to predict planetary motion exceptionally precisely. At first, Kepler's laws were widely ignored by astronomers like Galileo and Descartes, but when the laws were experimentally verified, Kepler's heliocentric model was eventually accepted, and half a century later they would provide a strong base for Newton to develop his theory of gravity.

Meanwhile, Galileo had built the first telescope with a powerful enough zoom to study the solar system through which he made numerous groundbreaking observations. Galileo claimed that the moon was not a perfect sphere, but instead, it was rough and uneven. He discovered that Jupiter, like Earth, also has moons and discovered four of them. Moreover, with his telescope, Galileo could spot all the phases of Venus, which, according to Aristotelian cosmology, was not possible. All of these observations strengthened Copernican cosmology, and astronomers eventually started transiting from the geocentric model to the heliocentric model of the universe. It was still, only after the publishing of Isaac Newton's *Principia Mathematica* in 1687, that a satisfactory model of the entire universe was formed, dominated by the idea of gravitation. Copernicus's heliocentric model, Kepler's law of planetary motion and Galileo's observations in astronomy were all expanded upon by Newton who resolved differences amongst previous cosmological models and also argued upon the infinite nature of the universe.

However, modern scientific cosmology is considered to have actually begun only by 1917. Albert Einstein's papers on the *"Cosmological Considerations of the General theory of relativity"* published in the journal *Annalen der Physik* were responsible for the advent of modern cosmology. Since gravity is the only significant force acting on cosmological scales, it was found that Einstein's field equations in general relativity not only provided a more accurate description of the universe than Newtonian Mechanics did but also set the basis of the theoretical understanding of the cosmos by integrating astronomy into physics to form a new science of *'cosmology'*. In his paper, Einstein had discovered a cosmological constant, which predicted that the universe was expanding further away by each passing day. This idea was revolutionary because earlier generations presumed that the universe was stagnant. And here comes a man in his 30's who has already toppled our view of time, length, energy, mass and even gravity saying 'and you thought you know the universe like the back of your hand.' In 1929, by studying experimental data on redshifts of starlight, Hubble discovered and firmly established the fact that the universe is actually spreading away by each passing second. He found that every observable galaxy is moving away from our own, and not only that, Hubble also found that farther a galaxy is from us, faster it is receding away.

Hubble's observations, probably the most important observational findings of the 20th century, were integral in the making of the model of the **Big Bang**, which was formulated throughout the 1930s. The concept of the big bang was extremely radical for physicists because the prior belief of a stagnant universe was hard to dispose of at that time, but the overwhelming evidence in favor of the big bang made them accept it. The idea is that at the beginning of the universe, there was a singularity that exploded due to hideous reasons and gave birth to the universe. But then this universe, expanding ever since its birth, must now be decelerating because the gravitational force of all the celestial bodies present in the cosmos must be pulling each other and slowing down everything down. This theory, while speculative, is the most popular theory on the evolution of the universe to date.

The unprecedented evolution of the universe can be divided into majorly six fragments. The *'inflationary era'* that preceded the big bang which was succeeded by the *'primordial soup era'*, which was then followed by the *'plasma era'*, *'dark ages era'*, *'stellar era'*, and lastly the *'dark energy era'*, the current era in which we live. The dark energy era is regarded as the last significant era of the universe, that started around six billion years ago and is expected to last until googol years into the future, till apocalypse of the universe. So, there is nothing much for cosmic historians to record as the significant events which define the universe's history have, unfortunately for them, already occurred. Fortunately for us, though, the gravity that emerged through the big bang was just the right amount. Had the density of matter a nanosecond after the bang been 0.00000000000000000001% greater, the universe would have diluted out to almost twice its size by current estimates, and our dear solar system would not have formed. Contrastingly, had the density been 0.00000000000000000001% lesser at that time, you would not be reading this line as the universe would have collapsed on itself only seconds after banging apart.

Now because the gravity turned out just fine, it is possible for us today to read and understand how exactly this miracle (the birth of our universe) actually turned out through history.

When It Banged!

Bring together each and every known particle from here to the edge of creation, absolutely everything, and compress it into a volume that is a million billion trillion zillion times smaller than the dot on this 'i', an unimaginably infinitesimally small volume that itself doesn't have any dimensions, this is also known as a singularity. Great. Traveling 13.8 billion years and one second back to a dark, stormy, and thunderous night, that singularity is what you see in front of you. Your spine shudders as you sense coarse rumblings anonymously erupt somewhere in the somewhere around you. Your goosebumps rise to the point they almost pull you out of your skin as you realize that you are here to witness a dramatic event — one which will result in a widely observed but unfortunately underrated term known as 'existence.'

This singularity is going to bang really hard. So hard that you definitely want to be at least a couple of tens of light-years away when it happens or else you will be overtly cooked before you even know that the bang has banged. But alas! There is nowhere to go. Before the universe began, it did not do so to fill any kind of void around it. The only space which existed was the one that was created as the material from the bang spread wider and farther into the unknown. Before the bang, there was nothing around around it, absolutely nowhere to go. If there ever was anything like nothingness, this is it. The only thing present was the singularity, waiting to unbecome one. But no, it could not wait. Time still had not come into being. Space and time were consequences of the bang. The singularity itself had no past. Furthermore, because there was no time no past, if you think about it, the bang happened just like that, out of the blue. Long story short, at time t=0, there was nothing. However, a billionth of a billionth second later, there was something, actually everything.

A unique incidence surmounting of an unperceivable amount of coincidences, a moment beyond glory or awe, a pulse too beautiful and expansive to put into words, our universe came into being. From never and nowhere, our universe became eternal and everywhere. Absurd, abstract, unimaginable, indescribable. You certainly do not want to be or even imagine yourself to be there.

Eras of the Universe

The idea discussed above is popularly known as the Big Bang. But, the theory of the big bang is misunderstood as one which tells us about cosmic origins. It does not. All it tells us is what happened just after the bang. In fact, the theory leaves us with many questions, including 'why' the bang occurred, 'how' did it happen, 'what' initiated it to happen. The inflationary theory of the inflationary era gives us the closest idea of what was there before the big bang. During the inflationary period, the universe had no matter, antimatter, dark matter, or radiation but a form of energy which was inherent to space itself. According to physicists, specific imbalances arose during the inflationary era, and this form of inherent energy caused that teeny-tiny amount of space which existed to expand exponentially with extreme rapidity. This very early expansion was so enormous that in every 10–30 seconds, the universe expanded to the size it currently occupies, i.e., approx. 46 billion light-years across. Essentially, the energy stretched the universe from whatever geometry it previously acquired to the one that Einstein's general relativity describes today. Moreover, during the big bang, for abruptly weird reasons which no one can figure out, this form of energy (inflationary energy) transformed into typical mass and energy that we witness today.

After the inflationary era, which initiated the bang, the primordial soup era dawned upon the nascent universe. This also marked the beginning of the four primordial forces – gravitational, electromagnetic, strong, and weak – which untangled themselves from each other trillionth of a second ensuing the bang. Moreover, the anonymous inflationary energy had by now transformed into known energy in the form of photons (massless pellets of light unable to decide their own identity between two simple options – a wave or a particle – consequently resulting in what we call today as the wave-particle duality).

A trillionth of a second has passed after the speculative bang until now, and space is filled with a soup of matter-antimatter particles, which included quarks, antiquarks, charged leptons, neutral leptons, and bosons. At this time, the cumulative heat of the universe is of the magnitude of 10^{28} Kelvin, enough for a photon to split into a matter-antimatter particle pair, which would, unfortunately, annihilate each other and become a photon once again. A game for the ages you may presume. Unfortunately, no. A millionth of a second after the bang, the universe had cooled down to about 10^{12} Kelvin, which was enough to stop this violent cooking. After that, all the matter and antimatter present, joined hands to annihilate each other while leaving behind unpaired particles in the ratio of around one matter particle in a billion matter-antimatter particle pairs. These loners who could not get married were responsible for all the matter we sense and perceive today as they had the fun of building the cosmos of the present. Had there not been this imbalance between matter-antimatter particles, the universe would have been filled with only photons – the ultimate *let-there-be-light scenario* as Neil deGrasse Tyson puts it.

Until now, exactly one second has passed after the big bang, and the universe is already a few light-years across. While the temperature is not hot enough to cook quarks and leptons, at 10^9 Kelvin, it is hot enough to cook electrons and positrons that go about playing hide and seek by annihilating each other and reappearing across the vast cosmos. Eventually, what was valid for quarks (the matter particles talked of above) is valid for electrons and positrons as well. One electron in a billion electron-positron pairs survives.

Around two minutes have passed until now. The quarks have rearranged themselves to form protons and neutrons, and the electrons are just the right number to nullify the positive charge of the protons across the fabric of spacetime. This results in a neutral but extremely dynamic universe filled with plasma, the fourth state of matter different from solid, liquid, and gas. Now, the protons and neutrons come together to form nuclei as the electrons still roam around freely. Naturally, it may occur to you that these electrons must come together with the positive nuclei to form atoms and give way for celestial objects to rise. Sadly, it was not that simple for these little fellows. As soon as the nuclei and electrons would come together to form an atom, an extremely energetic photon would come hurdling across space to knock them apart. There were just too many photons to allow the formation of neutral atoms, approximately a billion to one ratio.

What to do now? It is simple, wait.

But for how long? Around 380 million years, give or take a few million years. That is right. Only after approximately 380 million years, when the temperature of the universe had fallen from over 100 million Kelvin to under 3000 Kelvin, was the formation of neutral atoms actually possible. The energy of the photons was low enough to allow this formation as the universe came to the end of the plasma era.

Until now, whatever we have studied, is speculation through the theoretical understanding of how the universe's early days might have panned out. We do not have observational proof to ascertain if that really happened because for astronomers to figure out the mysteries of the universe, they should be able to see what's happening out there, and although we take for granted that the universe is transparent to light today, it was not the same always. During the plasma era, the photons traveled helter-skelter everywhere as they would strike and bounce off the randomly scurrying electrons to blur our view. This is a competing process because once a photon hits an electron in this plasma, it is absorbed by the electron and re-emitted a fraction of a second later in a variety of random ways (photoelectric effect). Only as the universe spread wider and wider, the energy of the photons redshifted to lower energies as the temperature too complemented by lowering down on the thermometer. Now once ordinary atoms were formed, after the end of the plasma era, the randomness in these subatomic particles reduced and these photons didn't have much to do other than report to astronomers and cosmologists 13.4 billion years later in an undistinguished part of the cosmos (Virgo Supercluster) in an undistinguished galaxy (Milky Way) near an undistinguished star (Our Sun) on a really well-distinguished planet (Earth), in the form of Microwave Background Radiation.

Now, as these ancient relics (photons) reach us today, you must think that astronomers must be able to see into our distant past and study the universe comfortably. The fact is that even if astronomers can study the universe through observational data until only 380 million years after the big bang, they cannot literally see what was happening in the universe at that time because these photons reach us in the form of microwaves which can only be detected through specially designed telescopes. Furthermore, only by using extremely sensitive equipment and after mapping out data for decades have cosmologists been able to study this radiation from the early universe that has since provided them with a wealth of information about our history. To literally 'observe' and look back into the initial stages of the universe's growth isn't that simple and through the telescopes we've built, we can see only until the end of the 'Dark Ages' when cosmic dust (majorly consisting of Hydrogen and Helium) had significantly parted away or had coalesced into heavier atoms.

Interestingly enough, if you ever want to see how the early universe actually looked like, all you have to do is to not tune-in into a specific radio or a television station. The data of every channel is sent to a radio or television through electromagnetic waves, and each channel works on a different frequency of waves. Hence, if you match the frequency of your device's receiver with that of a specific station, you can stream data directly from that station. If in case, the frequency of your device's receiver does not match with that of any station, then the antenna of the device starts picking up signals from the microwave background radiation, which is omnipresent across the cosmos. Therefore, whenever you do not tune-in into a specific television channel, you see a black and white image with many particles randomly moving across the screen in a very helter-skelter manner, that is nothing else but the blueprints of the early universe. What you see in front of you is, in fact, how the universe actually *looked like* during the plasma era. On the contrary, if you want to *listen* to how the universe sounded back then, then turn on your radio and set it to an unspecific radiofrequency, which is not associated with any channel. The radio's antenna will inevitably pick up sound signals from the pristine relics of the universe, microwave background radiation, and allow you to savor the ubiquitous ancient hums.

At the end of the 'Plasma Era' and beginning of the 'Dark Ages Era,' the gravitational force, at last, started exerting its influence by bringing together the free atoms roaming around space and forming the structure of the universe as we witness it today. Matter cooled and gravitated together to form stars, tens of billions of stars came together to form galaxies, many galaxies came together to form clusters of galaxies, and billions of these clusters came together to create the cosmos. Until around a billion years after the bang, the matter re-ionized while constructing the universe as the cosmic dust (neutral atoms ubiquitously present in the cosmos) faded out through major parts paving the way for light to travel unhindered in the future.

The end of this reionization process marked the end of the 'Dark Ages' as the 'Stellar Era' dawned upon the universe. The universe has now become transparent to starlight, allowing astronomers to zoom back, observe, and marvel at the first galaxies formed more than 13 billion years ago when the universe was still expanding, cooling, and interacting. Although the universe, at this point, continued to grow at incredible speeds, the rate of expansion started to drop as the billions of stars in the billions of galaxies began to use up and exhaust the free gas available to them. The galaxies started to merge while continually sucking up mater from their surrounding areas. The star-formation rate dropped, and as time went forward and the stellar death rate outpaced the stellar birth rate. By combining all of these and many more effects, one would naturally presume that the universe would start decelerating as gravitational forces would begin to dominate the vast landscape to result in a **Big Crunch** eventually. Unfortunately, it was not to be so as nature had other plans. Like an unexpected twist in the plot of a movie, our universe gives birth to dark energy, which marks the rise of the 'Dark Energy Era.' Dark energy acts as a force that is responsible for the objects (mostly galaxies and galaxy clusters) in the cosmos to accelerate away from each other. The dark energy era began roughly around nine billion years after the big bang and the expansion of the universe, cheers to dark energy, has been accelerating for nearly the past five billion years. It is projected that the dominance of dark energy over gravity would keep on increasing till oblivion as the universe, according to today's physics, will end up in the **Big Rip** where everything rips apart from each other due to this ever-increasing acceleration. Now, dark energy is aptly called so not because it is dark or black but because its nature even today remains oblivious to us. We have absolutely no idea what it is. We cannot categorize it or study its characteristics and leave alone to understand it we cannot even see it. Only after studying the anomalous behavior of galaxies and stars can we predict the presence of dark energy.

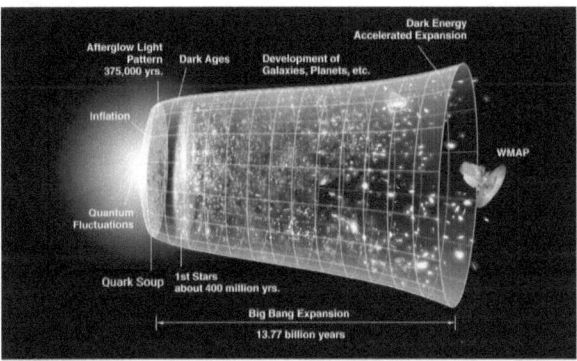

Figure 8.1 Evolution of the universe starting with the inflationary era, primordial soup era, plasma era, dark ages era, stellar era, and the dark energy era.

Historically, during the mapping of the universe, cosmologists found that they could spot only 5% of the theoretically predicted mass of the universe. Even after repeated efforts, the observable mass remained the same, and so did the theoretical mass. Were the equations wrong then? A definite 'NO'! The equations were crystal clear in their predictions and had been proven right time and again under various other circumstances. As it was later found out, it was our understanding of the complicated equations which was flawed. Einstein's equations predicted the presence of anonymous matter and energy, but because these ideas were too radical, no one believed them until it was vital. However, we cannot blame scientists for that. If, for example, your friend hits you and says, "it was an invisible hand which hit you, not me," you would be really foolish to buy that. It was a similar situation with the cosmos, but only true, where dark matter and dark energy combined to form this invisible hand. We will discuss dark matter and dark energy in detail in further chapters. As of today, no one really knows if the cosmos might surprise us tomorrow by flinging some other unexpected form of energy, matter, or god-knows-what substance in our platter that might force us to re-write decades or maybe even centuries worth of research and theoretical laws that might as well change the course and eventually the fate of our universe. However, that is precisely what scientific discovery is all about. Neil deGrasse Tyson says, *"the very nature of science is discoveries, and the best of discoveries are the ones you don't expect."* This vital speculation presents a strong point for imagination to the reader.

CHAPTER 09

Stellar Evolution – Dwarfs to Giants

"Every one of us is, in the cosmic perspective, precious. If a human disagrees with you, let him live. In a hundred billion galaxies, you will not find another."

– Carl Sagan

"Do not look at the stars as bright spots only. Try to take in the vastness of the universe."

– Maria Mitchell

"Not only is the universe stranger than we think, it is stranger than we can think."

– Werner Heisenberg

"The story so far: In the beginning, the Universe was created. This has made a lot of people very angry and been widely regarded as a bad move."

– Douglas Adams

Stars are undeniably one of the more critical components of the universe. Without them, our cosmos would pretty much be free-floating hydrogen as it was only their formation, which was responsible for streamlining and structuring of the universe. Their constant burning is responsible for forging heavier elements like iron at their very core, and their death is what makes it possible for planets to form. Stars come in mainly three sizes – red dwarf, average size star, and red supergiant. Red dwarfs are those stars who are around ten times smaller in size than our sun; average size stars are those who are around the size of our sun and at best ten times larger than it; rest of the stars which are greater than ten times the size of our sun are termed as red supergiants, and can grow up to a size 1000 times bigger than our sun. Specifically speaking, the diameter of our sun is close to 1,400,000 kilometers, but in extreme cases of red supergiants, we are looking at stars that are at least 1,000,000,000 kilometers long in diameter and even bigger.

Our existence today is a result of numerous deaths of primordial stars, which forged enough heavy elements for a planet like ours to be born. In essence, what you are reading is stardust, what you are listening to is stardust, what you are smelling is stardust, and in fact, you yourself are stardust. An atom in our head may be from one star, whereas an atom in your heart may be from some other star, who knows. But what is the life cycle of a star? Is every star the same? Why do they die, and how are they even born in the first place?

Let us have a look.

Birth and Life of a Star

Swimming in a molecular cloud (a cloud of Hydrogen gas) 100 light-years across, a miniscule Hydrogen particle not

Stellar Evolution – Dwarfs to Giants

more than 100 nanometers in diameter swishes through the bumpy, unorganized terrain as it smashes into its mirror image, another Hydrogen atom. In this unprecedented moment, when two Hydrogen atoms stick together, the combined gravity of this small pair just about makes the cut and attracts other Hydrogen atoms to join the fore. After millions of years of combining as more and more atoms come together, what used to be a single undistinguished atom, has now become an incredibly dense ball of gas. Its gravitational pressure is ever-increasing as more and more single atoms crush into this giant ball in a mad rush like never before forming what we call a Protostar.

The temperature in the center of this Protostar, after hundreds of millions of years, has risen well above ten million Kelvin, and in a moment of glory, something neat happens. The two Hydrogen atoms, which initially gave rise to this massive ball of gas, have now come so close together, thanks to the immense gravitational pressure, that the Coulombic forces of repulsion can no longer keep them apart and the Strong force overtakes. Fusion happens. These two Hydrogen atoms with a birthright of always repelling each other ignite to combine and form Helium in a misadventure clocking not more than a fraction of a second. Following this pairs' lead, many more atoms consequently begin igniting and combining. In this, not so mass-efficient process, energy is lost, which results in net outward pressure that is actually good. The inward pressure arising from gravity is now balanced by this outward pressure, and all the extra energy produced by this ignition is let out in the form of light and heat, the same light and heat which supports life on a planet like ours. Lo and behold, your star (just like the neighborhood sun) has been formed. This star is now a main sequence star where Hydrogen is fused to form Helium. Our sun is in this phase and has been here for more than four billion years.

Tons of energy is released every second as this giant ball of gas fuses Hydrogen atoms at its core while also keeping it away from imploding due to its gravity. As these atoms keep on producing more and more of Helium through fusion, Hydrogen completely depletes off at the core giving rise to a Helium core covered by outer layers of Hydrogen. This Helium core, on the other hand, shrinks and shrinks as pressure and temperature rise indiscriminately from the outside. As the temperature nears a 100 million Kelvin at the core, the Helium particles begin fusing into Carbon or Oxygen depending upon the number of particles participating in each fusion. Now, as this core becomes denser and denser, the fusion starts taking place faster and faster, and the radius of the star itself becomes larger as the energy released by the fusion gets bigger and bigger. At the point when the temperature of the core in the star rises above 100 million Kelvin, the star has already grown 100 times its average size and has turned into a red giant (our sun still has well over 4.5 billion years to reach that point).

Now, at this point of time when the star becomes a red giant, its Hydrogen supplies would start diminishing, and the star would only live on its Helium supplies fusing to form Carbon and Oxygen, but that supply too would soon be used up and the outer layers of gas around the star would wither away from this nuclear fireball to form a sort of planetary nebula leaving behind a hot dense core of Oxygen and Carbon. The star in this current state cannot fuse Carbon and Oxygen into further heavier particles because the temperature required to do so is well over 600 million Kelvin, which is unattainable. Hence, no more fusing takes place, and once the core itself stops fusing, the gravitational force would take over and the core would collapse on itself to a smaller, denser, and hotter sphere. This collapse is a sudden event and so powerful that it triggers a shockwave that spreads outwards and disperses the outer layers of gas of the star at incredible speeds, this phenomenon is also known as a supernova, which we will discuss later.

In the case of our sun, when the Hydrogen supplies would be finished, it would expand into a red giant and eventually absorb all of the inner planets until the core collapses on itself, and once the core collapses, it will trigger a shockwave expanding outwards in the fabric of space itself at light speed leaving behind the dense white core. This shockwave is predicted to consequently disperse the outer layers of gases at speeds of up to several kilometers per second, evaporating and absorbing any form of matter which comes in their path. In this process, elements are smashed together to form heavier elements, and the explosion of a star about the size of our sun will be able to fuse and form elements like iron, calcium, silicon, among others, until uranium and plutonium on the periodic table. This expansion of the outer layers of gas and formation of heavy elements in a mid-sized star, though, is not exactly a supernova but a more subtle variant of it. The giant cloud of gas moving away from the central core is termed as a planetary nebula but do not get misguided by its name as this place does not really give birth to planets, and the nebula itself fades out after some tens of thousands of years, a relatively minor period on cosmic scales. Eventually, the stardust of our sun would float around in the cosmos and end up in different living organisms on different plants in different solar systems after maybe millions or even billions of years. The dense white core, which was initially left behind after this expansion of gas, will become a white dwarf. This white dwarf star will just be and emit out its heat in the form of radiations as it gets cooler and cooler till the point it has released all of its energy. And at that point, it would become a black dwarf. Such a star would need more time than the age of our universe to actually form, therefore it is believed that no black dwarfs exist today. This is a lifecycle of a mid-sized star, a star which is nearly of the size and mass of our sun, and lives in its main sequence for about ten billion years.

Stars that are around ten times smaller than our sun are called red dwarfs, and according to recent cosmological models, these stars are expected to burn out their Hydrogen reserves in 6–12 trillion years and take another hundred billion years to cool down to become white dwarfs. Still smaller stars whose mass is lesser than 0.08 times the mass of our sun are classified as Brown dwarfs, and these stars are not even capable enough to fuse Hydrogen at their cores. Rolling about in the cosmos as a giant ball of gas, they are effectively incapable of achieving anything substantial in their lives.

Other stars, which are approximately more than ten solar masses heavy (mass of the sun), turn into red supergiants when their Hydrogen tank goes empty. These stars, due to immense pressure and heat, fuse Hydrogen at gigantically higher rates and monstrously hotter temperatures than our sun, and hence have a very short lifecycle spanning a million years at most. Once these stars finish their main sequence (use up all of their available Hydrogen), they start expanding. Then, once all of the helium fuel is used, these stars expand even more and begin using oxygen and carbon to produce elements like silicon. This process of expansion and fusion using heavier fuels depends upon the mass of the star and the temperatures reached by its core. Stars more than ten times massive than the sun can reach such immense sizes. In fact, the radius of one of the largest known red supergiant VY Canis Majoris, sometimes also referred to as a hypergiant, is about 1800 times larger than our sun. To put that in perspective, VY's size is equivalent to the size of Saturn's orbit. Talking about its lifespan, VY Canis Majoris can explode into a massive hypernova any moment now, and perhaps it exploded just now as you finish reading this sentence, but we will not know that for another 5000 years, which is when the light from the supernova will reach us. Generally, stars like VY Canis Majoris can reach temperatures of up to thousands of millions of Kelvin and usually fuse elements up to Iron at their cores. Though once the gravitational energy of their core becomes too high, the outward pressure rising from the fusion is unable to nullify the force and the star reaches instability. This results in a crashing collapse of the core that results in a massive shockwave, and supernova happens; in this case, it will be a hypernova because the size of the collapsing star is shockingly enormous.

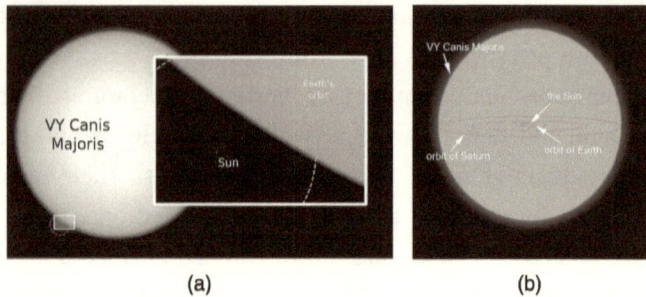

Figure 9.1 (a) The image compares the size of VY Canis Majoris and our sun on the best possible scaling. (b) The image compares the size VY Canis Majoris with the size of the radius of Saturn, which is only slightly larger than VY Canis Majoris.

During the explosion, matter particles are banged into each other with a tremendous force that is released from the shockwave. These furious collisions between matter particles result in fusion, and the atoms end up forming elements even more massive than the ones naturally found on Earth. Moreover, this explosion of gas from the supernova is so energetic that it more often than not outshines the host galaxy of the dying star itself. To put that into perspective, in 2015, ASASSN-15lh (name of a star situated in a distant galaxy 3.8 billion light-years away) outshined our sun, which is only eight light-minutes away, for 30 seconds when that star exploded in an unprecedented hypernova. Observations predict that ASASSN-15lh shined 570 billion times brighter than our sun and blew into one of the most massive supernova explosions ever recorded. Around 1000 years ago, in a much primary observation recorded in 1054 AD, a nearby star, now the Crab Nebula (shown in figure 9.2), situated around 6500 light-years away, exploded into a powerful supernova. The supernova's luminosity was so bright that the then Chinese astronomers recorded the phenomenon in history by explaining that they witnessed two suns in the sky parallel to each other for more than a week and this other star was described as a 'guest star.' Today, nevertheless, we know that it was not an alien star that visited the solar system for a week, it was, in fact, a supernova whose brightness was so immense that it seemed like there were two suns in the same solar system.

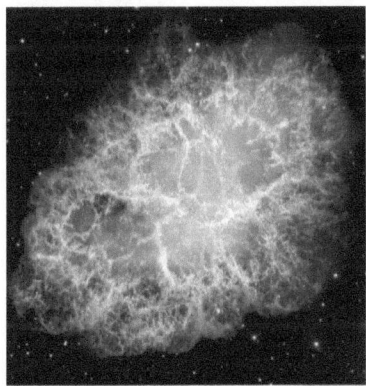

Figure 9.2 Crab Nebula. Remnant of the Kepler's star.

The remnant of Kepler's star, Crab nebula, today presents beautiful images of the birthplace of young stars. This cloud of gas is a product of the dead Kepler's star and is expanding at a rate of over 1000 kilometers per second even after 1000 years of exploding. It is already over 5.5 light-years across in size. A star of the size like that of the Kepler's star can explode into a supernova, unlike our sun, to form a nebula like the Crab nebula. This nebula hosts ionized Hydrogen, Helium, and other gases, which are essential in the formation of stars. There are regions in the nebula that are called molecular clouds where the initial formation of stars, as described at the beginning of this sub-chapter, takes place. As the young stars grow in size and mass over many years, they clear away their surrounding areas by ultraviolet radiation and make themselves visible to us back on Earth. The Eagle nebula, also known as the pillars of creation, shown in figure 9.3, is situated roughly 7000 light-years away and is a classic example of a star producing factory. The bright dots in the image are newborn stars hardly even a million years old, and all of them are born in the pillars of the Eagle nebula.

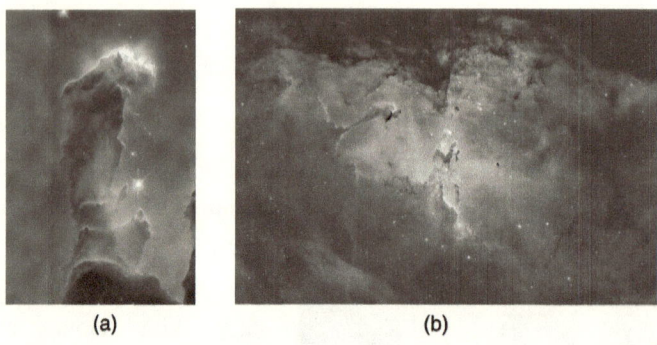

(a) (b)

Figure 9.3 (a) Tallest (Western) pillar of the Eagle Nebula. (b) Entire Eagle Nebula.

At 1300 light-years away, Orion Nebula, shown in figure 9.4, is one among many nebulae, that are solar system producing factories. Several stars in this nebula are forming planets through circular dust disks that have been captured by their gravitational influence. The trapezium star cluster inside the Orion nebula hosts many star systems which have circumstellar dust disks that are thought to be planets in the making, depicted in figure 9.5. Cosmologists have estimated that over millions of years, these dust disks would be frozen to form planets and stellar systems similar to our solar system. Still, the crab nebula remains to be one of the more studied astronomical objects in the night sky because it is a remnant of a red supergiant star and hosts a pulsar in its heart, a rare celestial object. Moreover, the nebula's close proximity to Earth and unparalleled beauty only adds to the fun. Essentially, the Crab nebula is an exceedingly attractive object of study for astronomers and cosmologists alike.

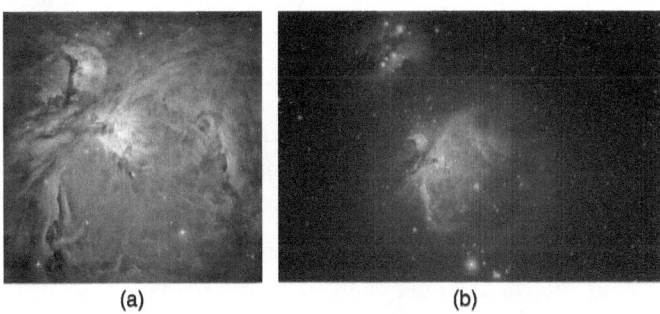

Figure 9.4 (a) A closeup of the Orion Nebula seen from the Hubble Space Telescope. (b) Orion Nebula seen from an Earth based telescope.

Figure 9.5 An artistic impression of a solar system in the making with dust rings orbiting the star to form planets. The Trapezium star cluster hosts many such solar systems.

A neutron star is an extremely dense ball of matter, which is the left-out core of a dead red-supergiant star. The explosion of the main-sequence star leaves behind such an extremely pressurized, dense, and hot core that the protons and electrons present merge to form neutrons. A typical neutron star is not more than 15 miles across (this is not a misprint). That is right; these stars are as big as a small city, and yet if not the deadliest, they are one the most lethal objects out there. A teaspoon of matter on this star can weigh more than a billion tons (as massive as a mountain), and their magnetic field is at least a trillion times stronger than the magnetic field here on Earth.

Still, many neutron stars remain undetectable because they do not radiate electromagnetic radiations but quietly emit x-rays while sitting in the center of supernova remnants. Some neutron stars are not so quiet and are constantly rotating around their axis under high magnetic fields. These small city-sized stars can rotate at a frequency of around 30 times per second, and the immense electromagnetic radiations which they emit can rip apart entire main-sequence stars. These small beasts are known as pulsars because of their periodic emission of ultra-high energy cosmic rays, which also makes them act as cosmic lighthouses. The magnetic field of a pulsar accelerates particles along the magnetic axis and produces funnel jets of matter that escape the pulsar. Through this regular emission, however, pulsars consistently lose their energy, and it is believed that they, on average, pulsate energy for about 10–100 million years before 'turning off' and going to sleep. In another type of a neutron star, called Magnetar, the magnetic energy is at least 1000 times stronger than that of a typical neutron star. The resulting energy bursts of these magnetars are so colossal that when in 2004 we recorded a starquake that occurred on a magnetar SGR 1806–20, it released more energy in one-tenth of a second than our sun has released over the past 150,000 years. Thankfully, this burst took place around 50,000 light-years away, had it been less than ten light-years away from us, it would have caused global extinction.

Stellar Evolution – Dwarfs to Giants

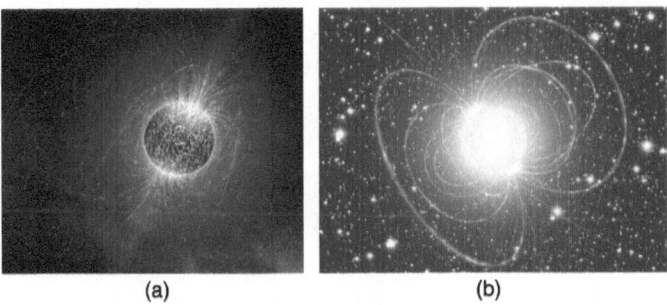

Figure 9.6 (a) An artistic impression of a pulsating neutron star, a pulsar. (b) An artistic impression of a Magnetar with its magnetic fields.

On the other hand, though, it is still not clear how pulsars and magnetars emit the powerful radiations that they do. Werner Becker of the Max Planck Institute for extraterrestrial Physics in 2006 said, *"The theory of how pulsars emit their radiation is still in its infancy, even after nearly forty years of work."* This statement reflects how incredibly difficult it is to study neutron stars.

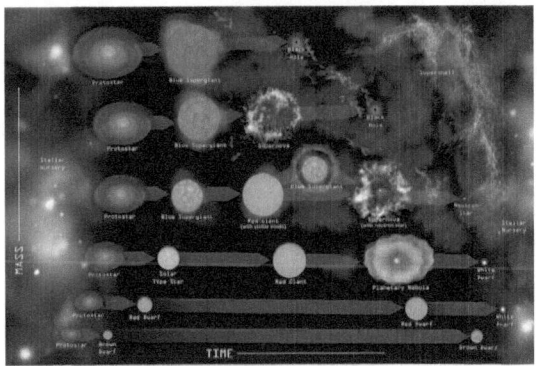

Figure 9.7 Stellar evolution cycle beginning from a stellar nursery and culminating into a stellar nursery.

The entire process of stellar evolution is visually depicted in the chart in figure 9.7. A stellar nursery is a molecular cloud where protostars are initially formed. Depending upon the size of the protostar, stars are consequently formed. A brown dwarf will always remain a brown dwarf. A red dwarf will burn for trillions of years to develop into a white dwarf. A mid-sized star, like our sun, will turn into a red giant then a planetary nebula and eventually into a white dwarf. Heavier protostars turn into blue supergiants (very early stages of massive stars), which subsequently turn into red supergiants. These massive giants then blow apart into a supernova, and depending upon the size of the cores of these supernovae, either neutron stars or black holes are formed. A supershell, or a superbubble as it is sometimes called, is another word for a molecular cloud that acts like a cosmic cavity hundreds of light-years across carved out from stellar winds and various cosmic activities. This supershell eventually turns in a stellar nursery to repeat this entire process again and again.

CHAPTER 10

Holes & Waves in the Dark Cosmos

"Black holes are where God divided by zero."

– Albert Einstein

"Who are we? We find that we live on an insignificant planet of a humdrum star lost in a galaxy tucked away in some forgotten corner of a universe in which there are far more galaxies than people."

– Carl Sagan

"There is a theory which states that if ever anyone discovers exactly what the Universe is for and why it is here, it will instantly disappear and be replaced by something even more bizarre and inexplicable. There is another theory which states that this has already happened."

– Douglas Adams

"Astronomers, like burglars and jazz musicians, operate best at night."

– Miles Kington

During the 3rd century BC, Hiero, king of Syracuse, appointed Archimedes to build and send a ship, Syracusia, 50 times bigger than any vessel built ever before, as a gift and a sign of power to his ally and the king of Egypt, Ptolemy III Euregetus. The ship was proposed to be built with inhouse hot water swimming pools, a library, eight watchtowers, a catapult to through 200-pound stone missiles, a temple for goddess Aphrodite, and a gymnasium. Moreover, Hiero intended to load the ship with 1000 men, 400 tons of grain, 400 tons of water and pickled fish, 600 tons of wool, and 20 horses. For people of that time, the task was equivalent to asking how to make a mountain fly. Moreover, to build a ship of that stature and then make it sink on its maiden voyage was undeniably not an option for Archimedes. The people asked, how is it possible? The Eureka moment is the answer. Sitting in his bathtub wondering how to make it possible, as history has it, Archimedes accidentally submerged a crown to realize the buoyant force of water and came up with what we call as the Archimedes principle today. Archimedes realized that because of a buoyant upward force, a liquid is displaced by a certain amount, which is equal to the weight of the object submerged in it. This principle was the reason for the success of Syracusia, popularly regarded as the Titanic of the past except for the sinking part, for which, thanks to our pal Archimedes.

Similarly, in 1605, when Copernicus, Kepler, and Galileo worked towards a heliocentric model of the universe, the church asked, but how is that possible? Gravity was the answer. Newton, as history has it, aggrieved at being hit by falling apples, decided to explain the unfortunate incident by coming up with the idea of gravity. And there you go, Copernicus's model, Kepler's laws, and Galileo's observations, which seemed convincingly improbable, were consequently not only explained theoretically but were well supported experimentally.

Even so, when in the 1820s it was discovered that Electricity and Magnetism are mutually dependent, people asked, but how is that possible? James Clerk Maxwell gave them the answer by developing his theory on Electromagnetism. He explained that electric and magnetic forces are a result of electromagnetic waves that travel in the presence of electromagnetic fields (refer to the first chapter for more information). Simply put, our universe is like a magician's hat where you never know what is coming next and when something new springs up, it leaves you in utter disbelief as you end up asking the same question once again – how is that possible?

Einstein's theory of relativity was no different. He was like an artist playing with a magician's wand developing something supremely beautiful but unimaginably far-fetched even to exist. The people asked, how is it possible? The discovery of black holes and more recently gravitational waves is a testimony to not only Einstein's theory, which was way ahead of its time but also to the fact that there is always an answer to the mysteries of the cosmos and nature decides to surprise you when you least expect it.

Black & White Holes, or Are They?

Black holes are cosmic beasts who have a free hand to torment everything that dares to block their path. When cores of massive stars collapse, they turn into neutron stars and pulsars. However, sometimes, when these cores are just too enormously massive (usually more than three times the mass of the sun), they collapse further under their gravity into what we call a singularity. All of the mass of the original star converges into this singularity, which is predicted to be not bigger than a mathematical point. This infinitely dense astronomical object gives rise to such immensely powerful gravitational forces that even light is incapable of escaping its clutches. And because no light particle is ever released from the surroundings of a singularity, a pitch-black circle is left behind, which looks like a gaping hole sucking in matter from anything in its vicinity and thus, justifying the black hole's name.

Just like stars, black holes also come in various sizes. Some are super-tiny, as big as atoms, and can be artificially produced in the Large Hadron Collider in Switzerland. While others are unimaginably supermassive and can weigh up to billions of solar masses heavy, the most massive black hole weighed yet is 40 billion solar masses heavy, which is located in the supergiant elliptical galaxy Holmberg 15a at a distance of around 700 million light-years away from Earth. These heavyweight champions are believed to be present at the center of most giant galaxies. Sagittarius A is the black hole at the center of our galaxy, Milky Way, and is recorded to be over four million solar masses heavy. It is not the most massive hole around but undoubtedly heavy enough to be classified as supermassive.

Unlike stars, though, a black hole's life depends directly on its size. We previously studied that larger and heavier the size of stars, sooner they die, but contrastingly, larger the size of black holes, the longer they live. Black holes, in general opinion, are supposed to swallow in material from their surroundings, which should imply that super-tiny black holes, formed inside the LHC in CERN, must grow in size at exponential rates by feasting on first the collider, then Switzerland, and eventually the entire Earth. Does this mean that we are in danger? Thanks to late Stephen Hawking, we are not. Hawking theoretically showed that Black Holes radiate out energy known as Hawking Radiations due to their thermodynamics. Super-tiny black holes, like the ones produced inside the LHC, radiate out energy faster than they can devour matter. So, as soon as these tiny holes are created, they start emitting out energy (information to be more precise), and before these tiny holes can take in matter from their surroundings, they themselves radiate out into oblivion. Hence, we can allow CERN to carry on with its experiments without fearing global extinction at the hands of a vampire star, black hole. Nevertheless, a black hole is still only an invisible creature of death and destruction; how are we able to spot it?

We can spot a black hole by not looking at its blackness but by looking at the matter that surrounds it. A black hole has immense gravitational energy, and it sucks the matter present in its vicinity, yet not all of the matter is sucked inside the black gaping hole, instead a lot of matter just spins around the center in a disc-like structure called the accretion disc due to the angular momentum produced by the deadly monster's rotation. The inner edge of this disc, which surrounds the black hole, is known as the event horizon, beyond which nothing (not even light) can escape the black hole's firm grip. Apart from spinning, a lot of matter is also thrown out of the accretion disc again due to the angular momentum of the spinning black hole, but the total matter in the accretion disc does not reduce because even more matter is sucked into it thanks to the black hole's powerful gravitational pull. In this furious cosmic dance of gas and matter around the black hole, magnetic fields are born, which heat the matter to monstrous temperatures of more than even a billion degrees at times. These hostile environments around black holes emit scores of radiations ranging from ultraviolet to microwaves, which are detected by various space and ground telescopes like the Hubble Telescope, Chandra X-ray Observatory, and W.M. Keck Observatory among others here on Earth. The data gathered by these telescopes and observatories help cosmologists to spot black holes, understand their behavior, and predict other features such as size, distance from Earth, mass, among others.

Another method to look for a black hole is by measuring extremely powerful gamma ray bursts or GRB. When a hypernova's core collapses into a black hole, which happens around once per Earth day in the entire universe, a gamma ray burst takes place. Why GRB's happen, we are not sure, but these explosions are *one of the most energetic events* in the universe where tremendous amounts of energy and gamma rays are released. Comparatively speaking, a GRB can emit more energy in ten seconds than the sun can emit throughout its lifetime of over ten billion years. An average GRB releases around $5 * 10^{45}$ Joules of energy, and thankfully for us, no such burst has ever been observed in the Milky Way. However, if an inevitable burst does take place in the future and that too in the direction of Earth, it will result in mass extinction. Optimistically speaking, a GRB is predicted to occur in Sagittarius A, the black hole in the center of our galaxy, only once in around 400 million years. Therefore, mass extinction on planet Earth by natural causes like gamma ray bursts is positively not as significant a threat as is global warming and climate change.

Figure 10.1 An artistic depiction of a gamma ray burst after a hypernova. The explosion of the gas in the centre of the image represents the hypernova while the linear jets originating from the same centre represent the direction of the burst of gamma rays.

 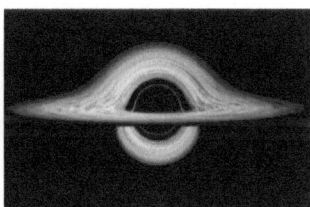

Figure 10.2 An artistic impression of a black hole spinning in reality.

Figure 10.3 An artistic impression of how a black hole will look like to us because of warps in spacetime.

The gravity of a black hole is so unimaginably immense that it can warp spacetime frighteningly nonchalantly. In figure 10.2, an artistic impression of a black hole shows how the accretion disc around the black hole should rotate in circles as naturally expected. But in figure 10.3, another artistic impression shows how we will see a black hole in reality if we are ever able to wander close enough to it. In this impression, the accretion discs that are vertically and horizontally spread around the black hole, are in fact, not two separate accretion discs but one single accretion disc. Due to the warping of spacetime in an awkward fashion, described by Einstein's equations of general relativity, the accretion disc, which should have been rotating in simple circular orbits, seems to be bent by a right angle. If observed closely, one will be able to clearly make out that the accretion disc does not break anywhere while it flows around the hole in a continuum in the fabric of spacetime. In the part of the picture towards the reader, the accretion disc is rotating in a horizontal direction around the hole, but as this accretion disc moves away from the reader in the picture, it slowly starts bending and seems to be vertically rotating around the black hole at the back end of the image. This visualization can be considered as an optical illusion produced by the fabric of spacetime that is under immense gravitational pressure created by the black hole.

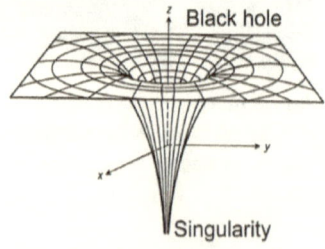

Figure 10.4 Depiction of the depression formed in spacetime due to a black hole's singularity.

When we talk about a black hole, we generally imagine an infinitely deep hole where matter is collected and should be collected till the end of time, but where is the hole in the black hole? The answer is that there is no hole. Figure 10.4 represents the gravitational effects of the black hole's singularity in spacetime but not its physical form. In reality, a black hole is not really a hole but a sphere. The singularity in its center is actually a mathematical point in the black sphere whose visible boundaries are marked by the event horizon. Beyond this event horizon, the gravitational pull of the black hole overpowers the escape velocity of photons, and consequently, nothing, except Hawking radiation, which is a different concept, can escape. In figure 10.5 (a), the central black circular region of the depicted black hole, in reality, looks more like the black metallic sphere shown in figure 10.5 (b). The bright circular disks, which can be spotted around the black hole in figure 10.5 (a), are accretion disks that consist of majorly dust clouds which are rapidly rotating around the black hole. Those who have seen the movie Interstellar must be able to relate to this description better. When the starship approaches the wormhole, Cooper realizes that what he was expecting to be a deep, circular hole was, in fact, a sphere. Cooper was then told that on paper we could draw a hole in 2-dimensions, but in reality, objects exist in 3-dimensions. Therefore, the circular wormhole in 2-D became a sphere in 3-D, which is what Cooper saw in front of him.

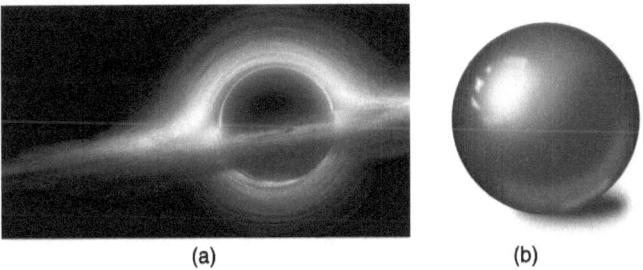

Figure 10.5 (a) Black hole as depicted in the movie Interstellar. (b) The central spherical region of a black hole inside the event horizon.

When Einstein brought out his theory of General Relativity, he could not imagine how anyone could solve the mess of equations he had produced. Fortunately for him, though, Karl Schwarzschild, serving in the German military, did come up with a solution that exactly solved his equations. But the solution predicted the existence of monstrous objects like black holes in the universe, and because these monsters were not observed until the 1970s, Karl's solution was not considered to be practically viable initially. But today, after we have confirmed the presence of black holes, Karl's solution is used to measure the black hole's radius through their experimentally determined mass. This radius is also known as the Schwarzschild radius and gives the value of the minimum radius of a sphere that the mass of an object needs to be compressed into to form a black hole. For Earth, the Schwarzschild radius is the size of a tennis ball. So, if Earth's entire mass can be somehow compressed into a tennis ball, Earth will become a black hole, and we may someday be able to live inside this black hole.

Apart from Karl's solution, which talks about a singularity and a black hole, Einstein, with his graduate student Nathan Rosen, in 1935, came up with a different solution to his field equations that predicted the existence of an interconnection between two black holes like entities. Einstein's initial purpose of drawing out another solution to his equations was to explain objects like black holes without using the concept of singularity. In his mission, Einstein eventually ended up building a cosmic bridge between two black holes, which popularly came to be known as the Einstein-Rosen bridge or a wormhole. This bridge has not been physically discovered to date but has prominent applications in sci-fi movies like Thor and Interstellar.

The basic idea behind a wormhole is that it connects two spatially separate points in spacetime in a straight-line path. We know that spacetime is curved and ever-expanding like a giant sphere. So, to travel between two points in this curved spacetime, one has to travel around the sphere on a curved path. Contrastingly, a wormhole provides an alternate route between the same two points in spacetime as it connects the two points internally. Shown in figure 10.6, is an artistic impression of a wormhole. Here, the curved part of the fabric of spacetime represents the actual path between two points in spacetime, and the tunnel-like structure joining the two points by a separate path is the wormhole. Hence, a wormhole is capable of reducing the travel time and distance between two points in spacetime as it provides a shorter distance between the two points than their actual distance in curved spacetime. Einstein's general relativity predicts that wormholes can reduce the distance between two spatially separate points from billions of light-years to only a couple of meters apart. Cosmologists even speculate that a wormhole may connect two different universes or even two different points in time in the same universe. These proposals nonetheless, remain as good as mere speculations because wormholes exist only on

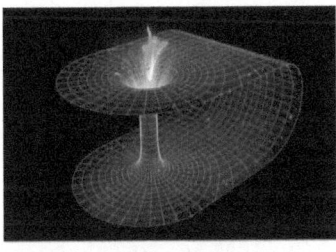

Figure 10.6 An artistic impression of a hypothetical spaceship entering a wormhole between two points in spacetime.

paper and not physically in the universe as per our current knowledge. Scientists have come up with brilliant ideas to form stable artificial wormholes capable of transporting humans and other objects through them. But these proposals are still only a fantasy as the materials used to form those holes are made of exotic matter, matter which is not believed to exist in the universe, for example, matter with negative density or a type of matter called non-baryonic matter.

Apart from black holes and wormholes, there is another set of solutions for Einstein's field equations called the white holes. These are an array of holes that are considered to hypothetically populate the cosmos because even though theoretically they are alive, these holes have not been physically observed yet. Consequently, not much is known about these holes due to a lack of observational data. But we know one thing for sure, if white holes do exist, they should behave exactly opposite to their black counterparts. Black holes do not let anything escape, but everything can enter inside them, contrastingly white holes should not let anything enter in as everything should be spewed out of them, probably even time. The Big Bang is believed to be the only credible white hole entity that ever existed in the universe, but there may be others. There have been several proposals to classify unexplained astrophysical objects as white holes such as GRB 060614, but none of the recommendations are convincing enough to be confirmed. Our understanding of black holes has grown immensely because of their physical sightings and the experimental data available for their study, unlike white holes. Had there been confirmed physical sightings of any white hole, we may have been able to progress on its physics. But as of today, the field of the study of white holes remains hollow, its future grim, and its only prospects remain in the hands of science fiction.

After studying all of these holes, three to be precise, we are still not finished with them. Sonic black holes, or dumb holes as they are commonly known as, are those objects which do not let phonons (sound particles) to escape from their vicinity just as black holes do not allow matter and light particles to escape. Interestingly though, sonic black holes are not some astrophysical objects which can be observed in deep space; instead, they are produced artificially in labs here on Earth. These sonic black holes occur in perfect liquids where the flow of the fluid is faster than the speed of the phonons, hence trapping the phonons inside the event horizon where the speed of the flow of liquid just equals the speed of sound also marking the boundary of the sonic black hole. These sonic holes are called *gravity analogs* because they are very similar to astrophysical black holes, and their properties are used to understand the properties of real cosmic black holes. Liquids such as Bose-Einstein condensates, one-dimensional degenerate Fermi gases, and superfluid Helium are used to produce these sonic black holes. It is crucial to note here that while sonic black holes are very similar to black holes to the extent that both of them even emit Hawking radiation, sonic black holes are not specific solutions to Einstein's field equations; instead, these holes were accidentally discovered during experiments. Hence, dumb holes, however smart they may be, are utterly unrelated to the general theory of relativity.

Ripples through the Fabric

"Although we will never be able to detect them, we must still believe that they do exist," such was Albert Einstein's faith in his mathematics when he predicted the existence of gravitational waves.

When you throw an object in a pond of water, the water immediately splashes upwards as the object comes in contact with the pond. This action gives rise to waves that originate on the surface of the pool, with the epicenter being the point of contact of the object with water. Contrastingly, when nothing travels above or inside the pool of water, the pool stands perfectly still without any disturbances. Similarly, when matter travels through spacetime, it also produces waves in the fabric of space itself, which undulate across the cosmos. Thus, the action of a supernova in space can be visualized like the action of large pebble tossed into a pond; the waves produced by the action of the pebble on the surface of water can be visualized as gravitational waves, but unlike any other waves which travel through space, gravitational waves travel within space as they are a form of space itself. Besides, these waves are called gravitational waves because they originate as a consequence of the depressions made by matter in spacetime, which is nothing else but gravity. Now because gravity is a curvature, as we have studied, gravitational waves are waves of curvature (depicted in figure 10.8). Hence, when gravitational waves travel across the cosmos, they cause the matter through which they travel to curve as well, but these resulting curvatures in matter are insignificantly minute.

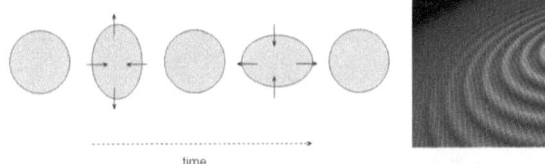

Figure 10.7 An exaggeration of distortion of Earth when Gravitational waves pass through it.

Figure 10.8 Computer simulation of Gravitational waves emerging from a system of neutron stars.

The concept of gravitational waves traveling in empty space seemed ridiculous to numerous scientists for many years because obviously, the idea of ripples traveling through something which is empty is mind-boggling. Hence, several scientists claimed that the mathematics of the waves itself was being misinterpreted. In due course, however, Einstein was proven right once again but only partially this time. The existence of gravitational waves was established, and as compelling the idea of an empty space maybe, it was confirmed that space, after all, can ripple. Now where Einstein got it wrong was the part where he said we could never actually detect them, because, in 2015, the LIGO Gravitational detector in the US did detect gravitational waves and confirmed their origin to the merging of two black holes in a black hole binary system some 1.3 billion light-years away from Earth. Now, because gravitational waves are distortions within space itself when they undulate, they stretch and compress matter according to the frequency of their peaks and troughs. So, during a trough, the wave and all the matter in it would stretch apart, and during a peak, the wave and the matter within it would get compressed and squeeze the same matter perpendicularly. The effect has been represented in figure 10.7 but is highly exaggerated only for the purpose of understanding.

When a gravitational wave passes through Earth, we can measure the distortions in matter using laser lights, which are extremely sensitive to even the most minute variations. The logic behind the experiment is that two light beams are shot between two mirrors placed at the ends of a machine, some four miles long, kept perpendicularly to each other, as shown in figure 10.9. The light beams travel back and forth in both the hands of the detector for an extended period and hence, travel millions of miles rather than just four miles. We know that a gravitational wave would stretch and compress spacetime, therefore when a gravitational wave passes through the detector, the laser beams traveling inside the detector would be stretched and compressed according to the geometry of the passing wave. Now, the gravitational wave could be traveling in any direction, and whichever direction in which it travels, it will comparatively produce more stretches and compressions in spacetime in one of the light beams than it will create in the other. The anomalies in readings hence produced between both the arms in the detector, will alert scientists that a gravitational

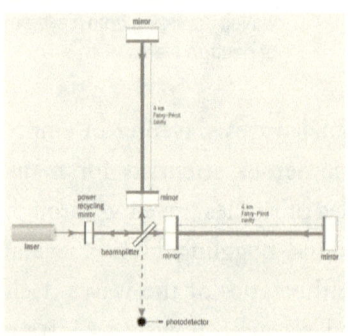

Figure 10.9 A general layout of a conventional gravitational wave detector.

wave has passed. In an exceptional case, there is a possibility that the gravitational wave passes through the detector at an angle of 45 degrees with both the arms. In such a case, the anomalies produced in the two arms of the detectors would be symmetrical, and consequently, scientists will not be alerted about the passing of any gravitational wave. These 45-degree angles are called *blind spots of the detector* because any wave passing through this direction will not be spotted by the detector. In 2017, gravitational waves were detected from a system of binary neutron stars merging for the first time, and interestingly, these waves were detected by only the LIGO gravitational detectors in the US but not in the VIRGO detector in Italy. This information was a very unexpected result because the VIRGO detector should have easily detected the gravitational wave and in each of the previous detections, both the detectors had made observations and provided similar results. Hence, cosmologists concluded that the wave must have passed through one of the blind spots of the VIRGO detector in Italy. This piece of information helped scientists to narrow down the search area and consequently pinpoint the location of the origin of these gravitational waves to a galaxy some 130 million light-years away from Earth.

The difficulty in detecting the waves is that they are incredibly feeble. The atomic bomb dropped on Hiroshima in 1945 packed a powerful punch of at least 15,000 tons of TNT, but even if anyone had stood right at the point of impact of the bomb with the ground, their body would have stretched by only as long as the diameter of a proton by the gravitational waves produced. It is a very insignificant amount of the effect of gravitational waves even when someone is so close to their origin. In reality, we try to measure gravitational waves emitted by astrophysical objects that are hundreds of millions of light-years away, which is equivalent to finding a needle in a haystack. Such measurements require extremely sensitive and highly advance technical equipment, and currently, only three detectors have been built in the world, one in the US (has two separate observatories – one in Washington and another in Louisiana), one in Italy, and one in Japan. The LIGO detector in the US, a collaboration between Caltech and MIT, is considered to be the most sensitive of all the detectors. There is another gravitational detector whose proposal has been accepted and is currently being built in India, named INDIGO, in collaboration with the LIGO detector in the US. This detector will enable scientists to survey a larger area of the cosmos by a magnitude of at least seven and thus help in expanding the study of gravitational waves by manifolds.

It is no shocker that the first gravitational wave was detected after 100 years of its theoretical prediction because the complex technical requirements needed to detect a gravitational wave, were hard to generate. Besides, this fact is evidence in itself to how far ahead we have come in theoretical physics when compared with the technological advancements of our generation. This revolutionary feat of the detection of gravitational waves is an epoch in human history. In the mid-seventeenth century, Galileo Galilei and Hans Lippershey opened the cosmos to our unaided eyes by developing the telescope. Their priceless work was integral to the evolution of today's telescopes and the progress we have made in the field of astronomy. During the mid-19th century, the discovery and extensive study of electromagnetic waves opened up new doors for us to understand nature. The crucial work done by scientists then has made it possible for the evolution of today's tech and the information-savvy world as electromagnetic waves form the backbone for communication; in fact, they form the backbone of our lifestyles. Similarly, the detection of gravitational waves has provided a fresh domain for scientists to view the cosmos. Instead of the electromagnetic force, in the years ahead, we will be able to use the ubiquitous gravitational force to tune in and literally listen to the secrets that the universe has for so long kept hidden from us.

The Most Tormenting Objects Yet

We have talked of nuclear bombs, supergiant stars, hypernova, neutron stars, pulsars, even black holes, and all of them are monstrously devastating beasts. What, you must think then, can be more powerful and devastating than that? Quasars. A quasar is a highly energetic jet of super-heated gas, which is accelerated to almost near-light speeds and ejected from a supermassive black hole (usually weighing billions of solar masses heavy) in the center of a massive host galaxy. Quasars are incredibly distant but the brightest objects ever observed. They are so bright that they can outshine our milky way from anywhere between 10 to 100,000 times. Moreover, they are so energetic that some of them spit out energy, which can be more than the energy produced by entire galaxies hosting trillions of stars.

Quasars were first discovered in the 1950s as sources of radio wave emissions from unknown astrophysical objects which had properties that defied logic and explanation. These sightings were named as quasi-stellar radios sources because of their nature and only later were they renamed to quasars. After the initial observations of the presence of quasars were published, it was widely proposed that quasars may be white holes emitting out matter into space. Granting that these proposals are quite popular and cannot be dismissed at once, a lot of definitive work is still required to prove that quasars may be white holes. Contrastingly, the physics behind quasars is not very detailed, and it is widely agreed upon that they are formed due to immense friction and pressure generated in the accretion discs of supermassive black holes. Physicists assume that these objects are formed only inside active galactic nuclei with a supermassive black hole, where an abundance of gas and matter from stars is available to feed the hole. The frictional force, which is generated when particles rub against each other in the accretion disc, emits heat and light, which is visible to us across the spectrum. As more matter is fed into the supermassive black hole, immense amounts of energy are produced, which escape the hole after streamlining and spiraling outwards above the event horizon. These massive, energetic, spiraling sprays of gas then stretch perpendicularly upwards for millions of light-years across space, and thanks to their brightness, they effortlessly dwarf their host galaxies. Refer to figure 10.10 to understand the occurrence of quasars through an artistic depiction.

Quasars have been citizens of the cosmos since the very early universe. The most distant quasar to be ever recorded is calculated to have been born when the universe was only 690 million years old. The peak epoch of quasar activity, as observed today, is recorded to have taken place around ten billion years ago, and cosmologists firmly believe that quasars are no longer active today, which may or may not be an accurate inference. Even our galaxy, the milky way, is presumed to have hosted a quasar probably around ten billion years ago when our black hole, Sagittarius A, was active. Today, however, our supermassive black hole is out of food in a stable inactive galactic nucleus, and hence it does not host a quasar. But when five billion years later our galaxy is predicted to crash into Andromeda Galaxy, our blackhole might just get the right conditions and ingredients to give birth to a new quasar all over again similar to the quasar Swift J1644+57 depicted in figure 10.10. The fact that the presence of quasars was much more common in the early universe can be explained based on the understanding that the galaxies then had much more dust and gas to feed to their black holes. Once the amount of matter fed into a black hole fell below a specific threshold limit, the black holes could no longer generate enough radiations for the survival of quasars. This finding also suggests that quasars played an integral role in galaxy formations in the very early universe, and hence, their study can provide us with a wealth of information to understand galactic evolutions.

Cosmic Reality

Figure 10.10 Quasar Swift J1644+57 was activated when a passing star near the black hole was violently sucked in. This doomed star's matter was highly ionized and consequently shot out of the black hole through a quasar jet.

These cosmic tornadoes, quasars, are the most potent objects known to us in the universe, but these objects are still not close to producing the energy produced in the most powerful cosmic events ever recorded. In this game of cosmic pinball, which we are a part of, merging of two black holes to form another black hole is considered to be the most powerful event in the entire cosmos. The energy produced by an average quasar is around 10^{44} Joules, which is equivalent to the energy produced by a million billion Soviet Hydrogen bombs detonating together. Whereas, the energy released in a cosmic merger of two black holes is at least a billion times greater than that produced by a quasar. The first-ever merging of two black holes was observed in 2016. Coincidentally, it was during the same observation that gravitational waves were detected for the very first time, which eventually helped scientists to calculate the energy released in this violent cosmic event to at least 10^{51} Joules during the peak of the merger.

The question now is, does this cosmic clash of titans cap the most energetic event in the universe? Perhaps not. We are still young and exploring and living through the dawn of the cosmic age. It is probably too soon to rush at conclusions when we surely do not entirely know what true powers may lay out there in the dark cosmos.

Why Is Everything so Dark?

In the early 1930s, an eccentric scientist, Fritz Zwicky, at Caltech discovered that a cluster of a few thousands of galaxies some 370 million light-years away was mockingly misbehaving. Theoretically, the cluster was supposed to throw away galaxies like water droplets from a spinning wheel, but Zwicky found out that the galaxies still remained clustered together. The mystery in his observations arose from the fact that the matter he could spot was incapable of producing an optimum amount of gravity to hold the galaxies together, but still, there they were. It was as if the galaxies were held together in a cosmic grip of matter which had no luminosity and hence could not be spotted. This unprecedented observation led him to propose the existence of nonluminous matter, which today is known as dark matter. Though, Zwicky's work included some uncertainties and was not entirely accepted as the skeptics of dark matter heavily outnumbered those who supported Zwicky's work. It was only in the late 1970s, when Vera Rubin and a team of other astronomers published a case study on the movement of stars in Andromeda Galaxy, that Zwicky's proposal of the existence of nonluminous matter was seriously considered. Rubin's work clearly showed that the amount of visible matter in Andromeda and other galaxies was just not enough to keep all of the stars clamped together. Vera concluded that her observations could only be explained by assuming the presence of some form of invisible matter that provided the extra gravity necessary to keep the stars in the galaxy held together. Rubin's work was extremely conclusive of the presence of dark matter and further studies in the topic consequently confirmed the existence of nonluminous matter termed as dark matter.

So, just like in Harry Potter when Harry wears his invisible robe we can't see him directly but can still see the effect of his actions in his surrounding like the footprints in the snow or when he beats up Malfoy and his accomplices outside the village, scientists cannot directly spot dark matter because of its non-luminosity but they can still predict its presence by observing the weird behavior of visible matter which is present around the nonluminous dark matter. Therefore, as it turns out, the matter you see, feel, and are yourself made of is actually free-flowing inside an ocean of dark matter that remains oblivious majorly because of our ignorance and partly because it does not react with known matter. In numbers, cosmologists have predicted that around a hundred billion dark matter particles (if their mass is calculated to be correct) are passing through your body each second without you noticing it, obviously. In general, according to calculations, the universe has five times more dark matter than ordinary matter. Now that we know that dark matter is present, we must then ask, what is it made up of?

Unfortunately, we have absolutely no clue. That is right, even though we know that dark matter is present somewhere out there, all we know about its nature is the substances it is *not* made up of, but what it is made up of, we are still working on that part. The reason we do not know what dark matter is made up of is that it does not react with the known matter at all. This mysterious noninteractive behavior of dark matter has also earned it the name of the ghost particle. For over three decades, scientists have valiantly searched for this ghost particle, and there is an increasing belief that we must be really close to deciphering its nature, but how close, no one knows. Detectors have been built hundreds of meters below the ground to make sure that undesirable particles do not interfere with the experiments that are attempting to detect the proposed dark matter particles. Furthermore, these detectors are built inside caves that are protected by thick layers of copper and lead, but, however strange it may seem, dark matter particles can still pass through all of these protective layers and hopefully one day, interact with a known matter particle by leaving behind conclusive evidence to establish its nature. The difficulty in detecting dark matter particles comes from the fact that it does not reflect light particles and nor does it absorb them. So, principally, the detectors themselves are clueless about what they are searching for eventually. The only thing that the ultra-sensitive detectors bank on is a theoretically predicted rise in temperature (namely a billionth of a degree of Celsius) when a single dark matter particle, in an entirely rare event, interacts with a known mater particle.

There have been reports of findings of particles interacting with known mater particles deep inside these detectors, but most of these supposedly unknown particles have turned out to be remnants of the early universe (cosmic microwave background radiation), which we already know are omnipresent. In some other cases, particles have presented robust cases for themselves as potential candidates for dark matter particles in addition to the various other theoretical proposals for the explanation of these particles, such as baryonic dark matter or even a vast swarm of miniscule black holes. However, none of these candidates and proposals have been concretely confirmed due to a lack of substantial evidence. After all, significant interactions between two known matter particles occur once in millions of billions of collisions that are performed in particle accelerators like the Large Hadron Collider (LHC) in Geneva, Switzerland, and when these collisions occur, they only last for about a billionth of a second, an unimaginably small amount of time. A collision between a known matter particle and a dark matter particle is an even rarer event that may occur for an even smaller amount of time. In simpler words, the entire process of detecting dark matter particles is an overwhelmingly long, tedious, and patience demanding process.

In chapter seven, what is so general about relativity, we read about warps in spacetime due to matter and energy. There we learned that light bends and travels in a geodesic path in the depressions made by matter and energy in the fabric of spacetime. This phenomenon is termed as gravitational lensing, and rightly so because it actually works as a lens and allows scientists to zoom deep into the universe.

Objects like galaxy clusters are massive entities of the universe, and are sometimes hosts to more than tens of thousands of galaxies; these massive occupants of the universe, quite understandably, do not allow astronomers to look past them as they block light coming from any object which is positioned right behind them. But, these clusters have unimaginable gravitational pull; they can warp the fabric of spacetime so much that light from galaxies and other objects which are behind them bends, and these objects are zoomed around the galaxy clusters in front of them. In figures 10.11 (a) and (b), the circular ring-like objects which can be spotted around galaxies are, in fact, galaxies themselves. These ring-like galaxies, also called Einstein rings, in reality, are elliptical and circular galaxies themselves, but because of the effect of gravitational lensing, they are stretched and zoomed around the galaxies and galaxy clusters in front of them. If you remember, we discussed this idea in chapter 7, *'what is so general about relativity,'* where we learned about the ability of matter to bend light.

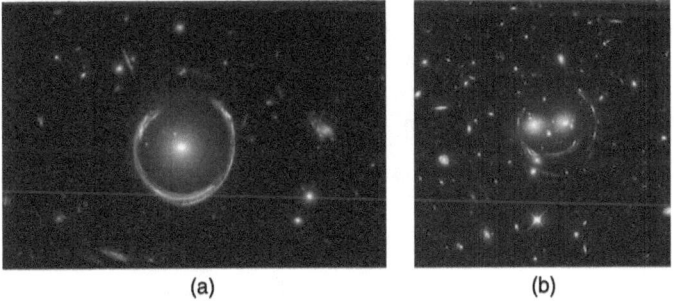

Figure 10.11 (a) A near perfect Einstein ring of a distorted distant galaxy looks like a horseshoe. (b) 'Smiley' or 'Cheshire Cat' image of galaxy cluster SDSS J1038+4849 produced by Gravitational lensing effects.

The effect of gravitational lensing is depicted in detail in figure 10.12. In the image, light from the background galaxy is bent around the foreground galaxy due to the depression created in spacetime by the foreground galaxy's gravitation. The undotted line, which connects the background galaxy with the telescopes on Earth, is a geodesic path. This geodesic path is the original path that light rays follow when they travel from that galaxy through spacetime. The dotted line shows how the telescopes on Earth perceive those light rays, and because of this optical illusion created by gravitation, a distorted ring-like image is produced of an originally circular galaxy. This effect is known as gravitational lensing effect, where massive galaxies act like cosmic lenses for us to observe and study cosmic objects which otherwise would remain oblivious to us. The gravitational lensing effect is not exclusive to only galaxies and cosmic objects because you can act as a gravitational lens yourself. The lensed light rays in that case, though, will be extremely insignificant to be observed because the gravitational pull from your body is negligible when compared to the gravitational pull produced by giant cosmic objects like galaxies and galaxy clusters.

Moreover, the gravitational lensing effect applies to not only visible light rays but any form of radiation. This feature allows scientists to better understand the gravitational lensing effect by cumulatively studying the data produced by these effects in different parts of the light spectrum. More importantly, as a consequence of this effect, telescopes can see galaxies, quasars, and other objects which they otherwise cannot see directly. There are mainly three different types of gravitational lensing – strong, weak, microlensing – which depends upon the size, mass, and position of the cosmic lens as well as the observer. The images represented above are examples of strong gravitational lensing, and such images are rare because galaxy clusters are not readily available. Other cosmic objects, which are not gravitational strong enough to produce such significant lensing effects, produce lower levels of lensing termed as weak gravitational lensing. Microlensing is a form of gravitational lensing in which the lensed light does not necessarily form multiple images of the background object; instead, the lensed image becomes brighter due to the extra bent light. This effect is helpful for astronomers to observe cosmic objects that are otherwise too dim to be visible.

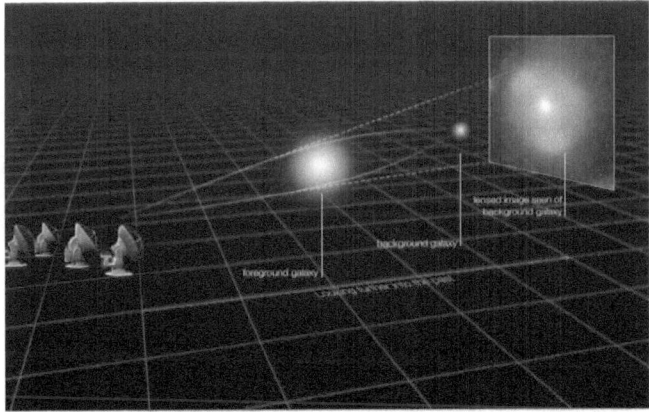

Figure 10.12 Light from a background galaxy is bent around the foreground galaxy because of its immense gravity. Telescopes on Earth observe this phenomenon as a lensed image like the one depicted in the figure. Gravitational lensing is nothing more than an optical illusion like refraction.

The phenomenon of gravitation lensing is specifically useful in not just viewing these lensed cosmic objects; instead, it has more profound applications in studying dark matter and its presence. Dark matter, as we know, is invisible but still influences its surroundings by exerting a gravitational force. Gravitational lensing effects, hence, occur due to dark matter as much as they occur due to ordinary visible matter. In many images of galaxies, cosmologists have recognized slight alignments in their movement that are intriguingly unexpected and according to cosmological theories, should not be there. These images, therefore, provide us with conclusive evidence of the presence of dark matter because only dark matter can be causing the unprovoked distortion and alignments in these images due to weak gravitational lensing that it produces. In general, we know that dark matter is omnipresent, this fact leads us to an exciting conclusion which states that most galaxies that we view are minutely distorted by dark matter, around 1% in most cases. Cosmologists can calculate the exact amount and distribution of dark matter in the neighborhood of these distorted galaxies by making minor assumptions based on concrete facts and applying the general theory of relativity on the data they collect from the images. Thus, the gravitational lensing of cosmic objects is a very reliable method for calculating the distribution of dark matter in the universe as the method relies on very few approximations to estimate the presence of dark matter in the universe, unlike several other less reliable methods.

Physicists make use of extremely powerful supercomputers and sophisticated algorithms to calculate and analyze how dark matter has influenced the growth of the universe after studying the distributions of dark matter. Some supercomputers, like the Summit supercomputer developed by IBM, can perform up to 200 quadrillion calculations per second. These supercomputers have produced simulations of the universe which show that dark matter has been and is present in the cosmos like an extensively interlinked cosmic web. The simulation shows that dark matter surrounds clusters of galaxies in a large sphere, and all of these clusters of galaxies are interconnected with filaments of dark matter, which extend from one galaxy cluster to another. This cosmic web of matter and dark matter, when compared with data from the Cosmic Microwave Background (CMB), has not only helped cosmologists to explain why the CMB is not uniform in temperature, but it has also helped them to understand how the universe grew from its infancy to its current size. In figure 10.13 (a), the brightest spots that you can see are galaxy clusters and galaxies made of ordinary known matter; the less bright and foggy filament-like structures which enclose these bright spots and are interconnected with each other are made of dark matter; all of the rest of the dark spots which can be spotted in the image are empty spaces where there is no presence of dark matter or ordinary matter, but only plain vacuum. The figure 10.13 (b), produced by a supercomputer for a small section of the universe, presents a timeline of how dark matter, which was randomly spread during the big bang, has streamlined to become as uniform as it can be observed in the present day after some 13.7 billion years of cosmic interaction and evolution.

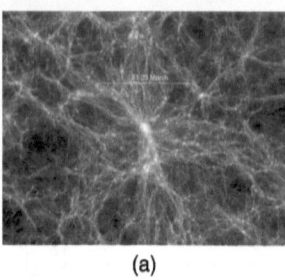 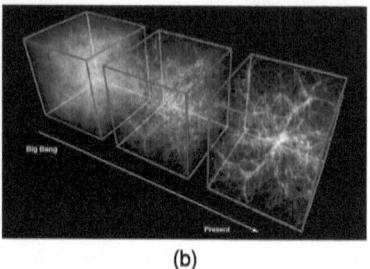

Figure 10.13 (a) A computer simulation shows the presence of dark matter around a galaxy cluster. (b) A computer simulation shows the evolution of dark matter through time in a section of the universe.

We previously read that Edwin Hubble, in 1930, discovered that the universe is accelerating away. Hubble's discovery gave birth to the concept of Big Bang, which seemed to be the only plausible theory to explain his observations. Dark matter, on the other hand, is known to bring matter together because of its gravity, and recent observations support this claim because instead of flinging apart, galaxies and stars are found to be clustered together across the universe because of the gravitational influence of dark matter. Therefore, astronomers used to believe that albeit the universe is moving apart, because of the big bang, it must still be decelerating because the gravitational forces of dark matter must be pulling back everything together. Hence, a team of astronomers, led by Saul Perlmutter at the Lawrence Berkley National Laboratory, set out to determine a decelerating coefficient, which would give them further information about the exact rate with which dark matter must be pulling the universe together. The method they employed to detect this constant was to study type 1a supernovae across the universe by measuring their brightness and gravitational redshifts of their emitted light. The reason they chose type 1a supernovae is that these supernovae have a dependable brightness (the core factor in determining a cosmic object's distance from Earth), which can be theoretically determined. Even though type 1a supernovae rarely occur (generally once in a thousand years in a typical galaxy), Saul's team was able to spot more than four dozen of them, which were enough to produce reliable data for their study.

After conducting their long and tedious experiment, Saul's team eventually did come up with a number, but that was not what they were expecting to find. Saul found out that until around seven billion years after the Big Bang, the universe was, quite understandably, moving apart but also decelerating at the same time. But some seven billion years after the bang, Saul found out that the universe suddenly and very abruptly started expanding apart and that too at an ever-increasing rate. So, what was supposed to be a deceleration constant eventually turned out to be an acceleration constant. This was a bizarre result. Ultimately, the whole scenario was taken back to the blackboard in an attempt to figure out what had gone wrong. But, even more bizarrely, Saul's team found nothing wrong in their experiment or the data it produced; instead, they discovered a new form of exotic energy which was dubbed dark energy. Saul's team found that this dark energy, which mysteriously sprung up seven billion years after the Big Bang, was reproducing more and more of itself from absolute empty space and behaved like a repulsive force. Saul's team realized that it was this repulsive force, dark energy, that had generated the acceleration constant they had discovered and was now responsible for pushing away everything in the cosmos while ripping it apart. Using the tools of physics, cosmologists have calculated that out of all the matter/energy inhabiting the universe, 71% exists in the form of dark energy, 24% in the form of dark matter, and only the remaining 5% is constituted by known visible matter. Essentially, this means that whatever we can visibly see, feel, and touch is not more than only 5% of everything which is actually present in the universe. And, paradoxically enough, we can explain the evolution of our universe, its working, and its features by using the two major components (constituting almost 95% of everything) that have never been seen, heard, felt, and of which we know nothing. This is not only strange but also an indication of the fact that how askew our perspective of the cosmos remains to be.

This discovery of dark energy brings us back to Einstein in 1917. Einstein's theory produced a cosmological constant (as Einstein called it), which predicted that the universe is accelerating apart at an ever-increasing rate. The amount of acceleration that Einstein predicted in his calculation, however, does not match with current estimates. This inaccuracy arises from the fact that Einstein believed in a static universe because, during his time, the idea of an accelerating universe appeared utterly ridiculous. Hence, he deliberately changed the value of his cosmological constant to fit his view of how the universe should behave rather than what the equations predict it to behave like. Later in his life, Einstein even admitted that the cosmological constant was the biggest blunder of his life as he unnecessarily fiddled with it. But the vital element in this story is that Einstein's theory was proven right once again after almost 80 years. Apart from this, the discovery of dark energy is significant because it is in accordance with the theoretical predictions of inflationary cosmology that explains how the big bang occurred and what happened afterward. Therefore, the results of the measurements made by Saul's team have not only established the existence of dark energy but in a way, the measurements have confirmed the authenticity of inflationary cosmology, which Brian Greene (a leading physicist) and others believe is the most important and lasting contribution of their generation to cosmological science.

On a separate note, the amount of darkness around us, 95% to be precise, is an alibi to the fact how much ahead we have come in our understanding of the universe since the Copernican revolution. For more than two millenniums, we had been told to believe that the Earth was the center of the universe; a little more than five centuries ago, the Copernican revolution changed that by making us realize that we are not the center of the universe, the sun is. During the second Copernican revolution we learned that the sun is not the center of the universe, the Milky Way is; even further down the line during the third Copernican revolution, we learned that the milky way is not the center of the universe, but the big bang must be the epicenter. Today, the dark revolution (or the fourth Copernican revolution as you may call it) asserts that we do not even know what the universe itself is made up of. Because what we know the universe is made of is only like a wreckage of a boat floating around in a river where the boat is the matter we observe, and the river is the dark matter and energy which the entire universe is made up of. On the same note, the idea of the existence of a multiverse, which is more prevalent today than ever before, is nothing else but the fifth Copernican revolution in the making.

CHAPTER 11

Time, Infinity and Beyond

"Science fiction is any idea that occurs in the head and doesn't exist yet, but soon will, and will change everything for everybody, and nothing will ever be the same again. As soon as you have an idea that changes some small part of the world you are writing science fiction. It is always the art of the possible, never the impossible."

– Ray Bradbury

"Time travel used to be thought of as just science fiction, but Einstein's general theory of relativity allows for the possibility that we could warp space-time so much that you could go off in a rocket and return before you set out."

– Stephen Hawking

"Two possibilities exist: either we are alone in the Universe, or we are not. Both are equally terrifying."

– Arthur C. Clarke

"The universe is under no obligation to make sense to you."

– Neil deGrasse Tyson

The journey we embarked upon since the time of Newton to explain the nature of space, time, and, in essence, reality has been a long one, universally agreed upon, though, is the fact that we are still only in the early ages of this long hauling journey. The evolution of the electromagnetic theory through the 19th century was ably exploited by the production of computer and radio networks around a century later. Half a century ago, no one could imagine a device that could give them information about all historical figures who had ever lived just on the press of a button, or the recipe of any dish they wished to prepare that too with video instructions while sitting on their living room sofa. But here we are today, where any institution or a place that does not host a wireless internet connection or even a computer network is considered obsolete; such has been our civilization's exponentially rapid growth.

The evolution of Quantum Mechanics through the 20th century, a theory integral to humanity's upliftment and not discussed in this book to avoid complexity and confusion, is today being practically exploited by our generation, which is on the verge of a revolutionary breakthrough to achieve quantum supremacy. Quantum computers, first hypothesized by Richard Feynman in 1982, are devices that work on the principles of Quantum Mechanics and are significantly more powerful and efficient than classical computers that are used in everyday life. In a leaked research paper in 2019, Google claimed to have performed a complex mathematical calculation by using its quantum computing platform in 200 seconds, which, for a classical computer, would have taken at least 10,000 years to complete. Such phenomenal is the generational improvement in technology from conventional to quantum computers! Some physicists even believe that by 2050, around 100 years after the quantum revolution, quantum computers will completely replace classical computers in the telecommunications industry.

Contrastingly, Einstein's theory of relativity, also developed through the 20th century, has still not been practically exploited to its full potential. Even after a century, we are trying to evaluate only its predictions. Science fiction, nevertheless, as always, is much ahead of reality and has employed Einstein's theory to good use. Traveling through space by using warp drives on the press of a button and time traveling to the past and future alike using futuristic time machines are only some of these sci-fi ideas, but the imperative question is, will we ever make these fantasies our reality? Having already failed earlier to predict how the information revolution will turn out and how the quantum computing revolution may turn out, it is wise for us to leave the speculation of future consequences of Einstein's theory to sci-fi writers only. Instead, in this chapter, we will discuss how far we have gone in our understanding of space and to some extent, the possibility of time travel.

Expansion of Spacetime

After the big bang occurred, we know that space and time were born as its consequences, and they expanded outwards at speeds greater than that of the speed of light, and as we have also read, things have not slowed down one bit since this early conception. Einstein's fundamental law, which sets the basis of his entire theory, states that nothing can go faster than the speed of light. Still, in conclusion, his theory says that the universe does expand faster than the speed of light. Nothing, you must think, can be more paradoxical.

The trick here is that in Einstein's theory, we talk about objects that are in motion relative to each other. So, if you are traveling near the speed of light, you are doing so relative to the cosmos, which is at rest, and the metric with which we calculate this relativeness is space, which forms the basis of everything. Think of it in this way, the ocean seabed provides the basis for water and life to accumulate over it, and it does not get affected by the actions that take place at the surface of the water. So, during tides, the water climbs and recedes on the shore regularly, but that does not affect the seabed. Instead, if the seafloor moves in the case of tectonic movement, it is the ocean waves that get disturbed, which may sometimes form a tsunami. Likewise, space is a tangible quantity that works like a skeleton for matter and energy to be present on, whereas it itself is not bound by the effects of matter that it supports. Phenomena, like gravitational waves and depressions in space, are also intrinsic properties of space, similar to tectonic movements of the seafloor that, at the end of the day, behaves only like a fabric. In the case of expansion of the universe, it is again space that expands but not matter that is responsible for that expansion. Now when space expands outwards, it naturally takes the matter present on it with it. This moving matter, however, moves symmetrically with each other (because of space) but does not move itself. To better understand this idea, study the following example.

Take an unused balloon and mark two points on it with a marker, now slowly and steadily blow air into the balloon and observe what happens to the two points; the two points would expand further and further away as the fabric of the balloon itself would expand. Now, on another balloon, mark two more points but this time mark them closer to each other and again fill air into the balloon; you will observe that although the two points move apart at a certain speed, this speed is still lower than the speed with which the points expanded in the first balloon. Now, in a third such balloon, again mark two points but mark them really far away from each other and fill air into the balloon; this time around, you will observe that the points would expand away from each other faster than in any of the previous two cases. The critical point here is that the points themselves do not move, but they seem to be moving at a particular speed relative to one another because the fabric of the balloon is itself stretching and expanding away. Contrastingly, if there had been only one point, you would have effortlessly concluded that the point is not moving at all as then there would not have been an external source to measure that expansion with. Similarly, space is also like a balloon, and as space expands, the fabric keeps on producing more and more of itself. Even more, unlike the balloon, which at some point will burst, space can expand infinitely forever as it keeps on stretching more and more without bursting apart.

Coming back to our analogy, think of the two points on the fabric of the balloon as two massive galaxies that are present on the fabric of space. Now had there been only one galaxy, an intelligent species like ours would conclude that nothing is moving away, but because there are two galaxies here, an intelligent species like ours observes that the two galaxies are moving away from each other and that too at a specific acceleration. Therefore, we perceive that a galaxy is moving away from us at a certain speed, and because of geometry, we also observe that farther away that galaxy is from us, the faster it is receding away from us and still accelerating away. In numbers, for every megaparsec of distance between two galaxies, the rate of recession increases by 71 kilometers per second. Except for the Andromeda galaxy, Messier galaxies, and a few others that are moving towards the Milky Way due to various reasons including gravitational attraction, most nearby galaxies and all faraway galaxies are receding away from us.

Considering the scenario mentioned above, a point comes in this cosmic attraction where the speed of the recession between two galaxies overcomes the speed of light because of this regular accelerated expansion. Hence, we can conclude that only space, which, after all, acts as a skeleton body for all matter and energy to reside on, has the permission to expand faster than the speed of light relative to the person who is observing that expansion. In reality, there are hundreds of billions of galaxies in the cosmos, and when we witness another galaxy from ours, we find that the other galaxy is moving away from us at a particular speed. But, if that same galaxy is observed by another third galaxy which is exceptionally close to it, the third galaxy would conclude that although they and the second galaxy are moving apart, the speed with which they are receding is lesser than the speed with which our galaxy and the second galaxy is receding away with. Similarly, when two galaxies are moving apart from each other at a rate higher than the speed of light, it is only relative to them; any other third galaxy near any of the first two galaxies would confirm that the speed of recession between them is actually much less than the speed of light. Hence, we can also conclude that it is only because of the intrinsic property of expansion of space that we observe galaxies to move away from us, where some of these galaxies are receding away even faster than the speed of light relative to each other.

How Big Is the Universe, Really?

The universe is 13.7 billion years old give or take a few million years. Thus, it must also be only 13.7 billion light-years far away in length in each direction. This simple understanding of the vastness of our universe is contextually flawed. It is a common trait of science that if you think too fast, it will end up making a fool of you. Take note that in the scenario of the universe only being 13.7 billion light-years in radius, we assume that it is static and has always been like that. But we know otherwise. The universe expanded colossally just after the bang, and due to dark energy, instead of decelerating, it has started accelerating away even faster. Therefore, the universe is much bigger than we think it is.

It is evident that the light rays we receive from the very early universe, in the form of Microwave Background Radiation (MBR), did not start their journey just yesterday. Instead, the light rays started their journey some 13.7 billion years ago. Moreover, we can also understand that

13.7 billion years ago, the universe was much smaller, and only because of expansion it is as big as we observe it to be today. That is to say that the distance between the source of light rays (MBR) from the very early universe and our Earth was only around 42 million light-years when the rays first started their journey. But due to the consistent expansion of space, the distance itself stretched apart, and the light rays had to journey for over 13.7 billion light-years to cover a distance, which was initially only 42 million light-years in length. Even more, the origins of the light rays themselves have expanded away, and according to current calculations, the boundaries of the visible universe, marked by the farthest objects that we can see in any direction, lie around 46 billion light-years away from us in each direction. Hence, the diameter of the observable universe, which is the part of the universe we can directly observe from Earth, is around 92 billion light-years. Explicitly speaking, photons from objects beyond this 92 billion light-years extended sphere, cannot reach us because those objects are moving away from us at a rate much faster than the speed of light. Hence, we may never be able to witness what lies beyond this far edge; a portion also referred to as the unobservable universe. Furthermore, this also implies that the most distant galaxies which we can witness today, some that are more than 13 billion light-years away, will not be visible to us a couple of centuries or millenniums later. Principally speaking, it is better to study and document the presence of these objects now than never, because we have already missed out on much information that has been flung out of the realms of our farthest optical reaches.

Time, Infinity and Beyond 201

Figure 11.1 This is a Hubble deep field image. Bright circular spots and point like objects whose light is scattered are stars within our galaxy, many other galaxies, which can be made out in the image, are many millions of light years away. All the rest of the tiny speckles that can be spotted in the image are galaxies which are billions of light years away, the farthest studied being more than 13.2 billion light years old. This image will not remain the same in the future when some of these speckles would be flung out of the observable universe.

Remember a time you were standing by the side of a railway track or the road and heard a train or a car zoom past you. When the vehicle approaches your position, you hear that the sound emanating from its movement congregates and becomes heavier and heavier, but once the vehicle has passed your position, suddenly the same sound becomes shrill, diluted, decreases in magnitude, and consequently fades away. Similarly, light is also a wave and when space itself expands apart, the light wave coming from a point in space which is moving away, stretches out in length and magnitude, and once the expansion rate becomes larger than the speed of light, these same light waves are no longer capable enough to reach their destination, which is us. So, even though matter emits light and radiation beyond the observable universe, those radiations and photons will never be able to reach us thanks to the expansion of space at such a colossal rate. Theoretically, however, we can still estimate how big the universe may be if we may not know exactly how big it is. In fact, an estimate suggests that *the actual universe is as big to the observable universe as the observable universe is to the size of an atom.*

The reality of the size of the universe also explains why the sky looks dark at night. We know that there are billions of trillions of Herculean stars shining brightly out there in the cold. Their brilliance should quite naturally flood the night sky with light, but that does not happen. This happens because we receive light from an estimated only 0.00000000000…00000001% of all the predicted matter which should theoretically be present out there. Therefore, because the light rays beyond the observable universe have not reached us yet, the sky is not lit up entirely. If someday, all of that light accidentally does reach us, the night sky would be so bright that it would blind us out (a physics breaking day it would be when that happens).

Nevertheless, what is/are those things that actually lie beyond the visible cosmic boundary that nature has set for us? The suggestions are innumerable, but we must explore some, which are strongly influencing and seemingly true to their core.

A pretty apparent first suggestion comes out to be that the universe must be infinite. I mean, what do you expect, a certain kind of a cosmic wall would suddenly appear at the edge of the universe and stop it from expanding, the idea of the wall is credibly ridiculous; therefore, we must not entertain it. Therefore, as an alternative, the universe must be infinite; if it is not, then it indeed is expanding into one. Now, when we talk about infinity, then anything is possible. Absolutely anything. Some trillions of light-years away from your home right now there might be another you who surprisingly enough might be reading this exact same sentence in this precise same book with the only difference being that the person might have picked his/her nose while reading and you did not, or maybe you did. Besides, there might be some other you far away who might be in jail for assassinating the POTUS and some other you might be standing in some other Stockholm receiving a Nobel prize in Physics while you sit here and envy that person. The principal point is that – Anything. Is. Possible.

But, if in case there is a particular cosmic wall, then what happens? Fortunately, our physicists do have another suggestion to explain the universe in that scenario. To counter the idea of a cosmic wall, our physicists propose the existence of a multiverse and declare that our universe is only like a bubble in a vast ocean of universes. If accidentally, our universe strikes with another bubble universe, the other universe would supposedly act as a wall; the effect is depicted in figure 11.2 (b). Moreover, when these two bubbles collide, then so must their gravity and physics, and who knows if the other universe may have some other laws of physics than ours. A collision of these two physically inconsistent universes, physicists believe, will result in a dissolution of both of them and then consequently result in another big bang to give birth to a third universe that might even be compatible with the laws of physics of the first two universes included. Such a suggestion may be quite persuasive at first glance, but is still food for science fiction as it is merely a speculation and we have no substantial proof to even confirm the existence of a bubble multiverse.

A third suggestion is that of the Big Bounce, the idea of a *cyclic universe*. Some physicists believe that the current state of the universe is headed towards oblivion, where dark energy would become even stronger in the future and tear apart everything that is there to be seen. The expansion of space would become so enormous that leave only atoms even spacetime would be spread apart so much that nothing would remain as cosmic nothingness would eventually reign. This speculative idea is referred to as the Big Rip. But the story does not end here. Physicists speculate that in this cosmic nothingness that our universe has turned into, nothing would suddenly expand into everything once again through another Big Bang. The new universe would then go on to keep expanding for some billions or trillions of years and eventually turn into cosmic oblivion once again. This cycle, physicists believe, will repeat until forever. The interesting point to consider here is that we do not even know if we are the first, the last, or in the middle of this cosmic cycle of universe regeneration. Essentially, the theory says that reality is like layers of universe stacked one after the other. It is similar to the rings of Saturn, where each ring represents a separate universe. The only difference is that in our reality, the rings are probably much higher than nine and evidently indefinitely vast.

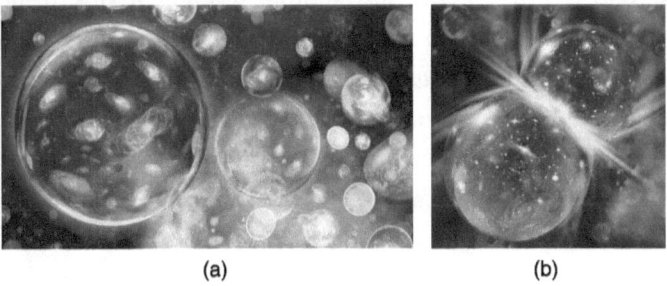

Figure 11.2 (a) An artistic impression of a bubble multiverse. (b) An artistic impression of a collision between two universes in a bubble multiverse.

Time is a funny thing, and the idea of the Big Bounce suggests it may be possible that we are prisoners of an eternal time frame repeating the same process over and over again, similar to how Stephen Strange imprisoned Dormomu in a time loop in the movie Doctor Strange. After being born from nothing to diluting into nothing and repeating the same cycle, the story of the life of our universe is undoubtedly the greatest ever told.

Time Travel

A harsh truth about time is that no matter what we do or anticipate about it, it always runs in one and only one direction, forward. Call it old-school, undemocratic, or totalitarian of time to not allow us any choice of whim to travel through it as per our liking, but that is just how it is, and sadly, we have to live with this fundamental law of the universe. Contrastingly, when we travel through space, we have the power of choice to decide if we want to go here or there, something we lack in the domain of time where every second is succeeded by just another second. Brian Greene says, *"Were we able to navigate time as easily as we navigate space, our worldview would not just change, it would undergo the single most dramatic shift in the history of our species."* An irony here is that albeit we cannot traverse time as easily as we traverse space, for almost over a century, we have known how to travel forward in time if that is all that we are allowed to do.

Einstein, through his Special Theory of Relativity, provided us with a blueprint for time-traveling into the future in 1905. If Mahesh boards an intergalactic spaceship and leaves full-throttle at speed say 99.99999 percent the speed of light for Andromeda galaxy, on his return, he would observe that millions of years would have passed on Earth while he would have himself aged only a day older. In this way, Mahesh would end up traveling into the future. This is an undisputed and theoretically robust claim which has been experimentally verified time and again. The only shortcoming in this otherwise foolproof method to travel into the future is the lack of a spaceship that is capable of traveling at 99.99999 percent times the speed of light. The fastest human-made object, Helios-2, a sun orbiting probe, recorded a mind-boggling speed of over 210,000 kilometers per hour while orbiting the sun. This speed, nonetheless, is still dwarfed by the speed of light. The prospects of humanity making an object capable enough to travel at near light speeds in the near future are grim. Furthermore, even if we can build a ship with that sort of capability in the very far future, it is hard even to speculate how that spaceship might be made to work, the technological advancement required

Time, Infinity and Beyond

for building such a spacecraft by current estimates is too far-fetched to be achieved.

The issue of traveling into the future, you may think now, is more complex an endeavor than previously thought. Dodgier issues arise, however, if we try to work out the possibility of a way to travel to the past even if only in our imaginations. The grandfather paradox is one of the more popular ideas to explain this dilemma. The paradox says, what if you could travel back in time and kill your grandfather before he could have kids. This murderous intent, not well received by your parents, results in the fact that because your grandfather never had kids, you must never have been born. Now because you were never born in the first place, the paradox argues, how did you kill your grandfather in second place. And even if we believe that you did kill your grandfather, that would mean that you will never be born, which would mean that your grandfather will never be murdered; now that your grandfather will never be murdered, this will ensure that you are born, and, as we all know the malicious intent you have, you will ensure that your grandfather will be dead before he has kids. Subsequently, the entire loop that we have just pondered upon will run all over again. Essentially, this means that your grandfather will be dead and alive at the same time, and we, time travel police, will have no way to confirm if he is either of the two at a given point because if he is dead, he should be alive, and if he is alive, he should clearly be dead. This is a logically illogical impasse that we have so uninvitedly trespassed.

Another similar paradox, first proposed by Oxford philosopher Michael Dummett and his colleague David

Deutsch, provides us with more reason to believe that time travel to the past, if possible, must have complicated restrictions of its own. Let us assume that I build a time machine today and travel fifteen years into the future. In the future, I learn that my father, absolutely uninterested in physics and cosmology as I knew him, is world-famous for publishing a paper that describes an effective way for us to build an intergalactic spaceship that can travel at near light speeds. While reading his groundbreaking yet modestly simple research paper, I come across the section of acknowledgment where my father has thanked me for the years of instructions that I had given him in physics that today had helped him in writing his paper. This unexpected discovery leaves me bamboozled and I decide to go back into the past to start teaching my dad. So, back I come from the future and start training my dad in high-energy physics and aerospace engineering, which I believe will be enough for him to come up with his groundbreaking revelation. A year passes. Two years pass. Five years, ten years pass, but I still do not see any sign of my father understanding even a bit of the esoteric ideas that I have been sharing with him throughout. I start becoming impatient and hasten my lessons. Another four years pass, and my concerns are sky high as the progress we make is damply minimal. One year before the scheduled publishing of the paper, I unwillingly realize that my dad is not going to discover his idea on his own; instead it has to be me who will have to dictate him the entire paper word to word as I read it in the future. And so, I do. Eventually, my father publishes his paper right on time, and the events, quite unexpectedly, turn out exactly as I had seen them in the future fifteen years ago. The only missing piece in the puzzle, as it turns out, is that where did the information and the ideas published in the paper actually come from? My father certainly did not come up with them, which he logically should have, and I did not come up with the ideas either because I read them in my father's paper by going into the future. In reality, I cannot point to any person or device that can be credited for coming up with the ideas which my dad is being credited for. In essence, I conclude with concrete evidence, that in the world where time travel to the past and the future is equally permissible, knowledge and information must materialize out of thin air. No piece of information, I assure you, can be weirder than this in your entire day today.

In conclusion, what should we conclude from the intimidating paradoxes which strongly suggest that time travel to the past is not only filled with nonretractable complexities but also results in terrifying consequences? The truth is that we must not draw conclusions based solely on puzzles that we have merely imagined and not practically verified in real life. In fact, if this may console you, we have come up with other ideas that logically and mathematically make time travel into the past theoretically permissible.

Timeless Time

Time has classically been considered as a river that flows on and on with no returning. This is an appropriate method to envision the flow of time, but to better understand how time works, we must think of it as divided into moments. Take a big loaf of bread and attach a meter scale to it. Slice the entire loaf into paper-like thin slices without disturbing the bread. You will observe that each slice is categorically different from the other based on its position in space, which can be determined by the meter scale. Likewise, time can be treated like a loaf of bread where the slices of the loaf are photographic images of moments with each of them being categorically different from the other based on its sole time coordinate. Fascinatingly, when all of these moments are strung together, time seems to flow just like a motion picture, which is an accumulation of photographic images, and just like photographs, no individual piece of memory can come alive on its own. In essence, these unique moments are timeless, and each of them acts only as a memory; these moments cannot substantiate into the flow of time individually. These timeless moments are just there, behaving like indestructible and unchangeable slices while being stowed away in a cosmic hard drive of time (the term *'cosmic hard drive of time'* is an assumed term that is used in the book to refer to the concept of flow of time which is secure in such a way as if it were stored in a hard drive that can't be meddled with).

The realization of how time actually works leads us to make some exotic conclusions. We have all, at some time, nurtured the fantasy of traveling through time with the hope of changing our pasts at will. But now, we must accept the reality that ensued moments in time are, after all, *unalterable*. Whatever has happened, has happened and cannot be changed. In the grandfather paradox, you going back in time with the murderous intention of killing your grandfather is nothing new for the cosmos. Let us say you go back to 4:00 pm GMT on 31st December 1968 in your granddad's parental home where he grew up as a kid. This act of you going to your grandfather's house is a memory well-distinguished in the indestructible loaf of moments because the three coordinates of space and time are unique in a plethora of memories which the loaf contains. So, whenever anyone looks at the slice of the loaf that holds the memory of you visiting your grandad's parental home at 4:00 pm GMT, 31st December 1968, that person *will always* see you there as you will *always be there*. You were *never not there* in those four unique co-ordinates of spacetime. The curious fact is that whosoever searches for you at 3:59 pm GMT, 31st December 1968, at your granddad's parental home, that person will *never* find you in that slice of memory, but precisely one minute later at 4:00 pm GMT, you will *spring out of thin air* claiming to be from the future. So, the act of you embarking on the journey to kill your granddad was *always bound to happen* as the turn of events were already predicted and stored in the loaf of moments kept in the indestructible cosmic hard drive of time. Hence, your presence in your grandfather's home on 31st December 4:00 pm is an *eternal* feature of spacetime, like an irritating unremovable watermark.

This unanticipated form of the flow of time, which asserts that you will always be there to kill your granddad, now brings our attention to the next fact, is it really possible for you to execute your grandfather after firmly deciding that you will do so even though you know that moments in time are immutable and can't tampered with? The thought-provoking answer is that you will never be able to succeed on the mission you so excitedly embark upon in the first place. History is an alibi to the fact that your grandfather was not murdered by a ridiculous man who claimed to be from the future because you were born. So, it will remain like that. This is an event of time which cannot be changed. Hence, no matter what you do or how much you try, you will never be able to kill your grandfather. What specific reason stops you from killing your granddad is not physics' job to predict, but it does predict that because you are born, you must not be able to execute your grandfather before your father was born. On a separate note, speculating upon the reason for your failure to execute your grandfather, maybe your mind changes when you see your granddad as a kid, or perhaps you never find an appropriate moment to attack, or perhaps some fancy dress people claiming to be time police materialize out of thin air to arrest you on the grounds of malign intentions and misuse of time machine (they read your mind using a mind-reading device from the future). We will not know for sure what may cause your unsuccess until someone someday practically attempts to carry out the paradox in its entirety, which, if possible, can take place only in the very far future; you may not live to see that glorious day. In essence, something or someone (maybe a messenger sent by God) will always be there to stop you from succeeding in your ethically wrong and ridiculous mission (I mean, who wants to kill their grandfather in such a way).

Conclusively, this new method of the depiction of time disrupts the possibility of meddling with your past because it presents us with a cohesive set of moments that *have to take place*. These moments have to be carried out into events in the same order in which they are aligned. You developing the idea of killing your granddad and going back in time to fulfill this thought-to-be prophecy is just a small part of the bigger picture that was supposed to take place. You were just helping the play of time by fulfilling your own destiny. Ultimately, the entire idea elucidates that time does not present a contradictory story, if rightly interpreted, time gives us a sensible and logically coherent set of events that, terrifyingly enough, cannot be altered or meddled with. The entire concept of unchangeable time is well portrayed in a statement by Brian Greene, where he says, *"If you time-travel to the past, you can't change it any more than you can change the value of pi. If you travel to the past, you are, will be, and always were part of the past, the very same past that leads you to travel to it."*

This analysis, which shows that although we can travel to the past, we cannot change it, comes from a logical understanding of how time works according to modern-day physics. That being said, our knowledge of time travel just gets more streamlined by this analysis as it circumvents the hindrance offered by ambiguous situations, which could possibly disallow any potential prospect of time travel to the past. On the other hand, the ridiculous rule which states that you will never succeed in such a mission as executing your grandfather can be better understood only when it is practically exercised in a real-life situation. The reason it is so tough to grasp this rule is that it violates the fundamental notion of our ability to practice free will.

The Past and Gödel

Physics is a very deterministic subject as it allows you to predict the future of objects and events based on their current state and properties. If you know the quantum wavefunction (quantum wavefunction encodes all necessary information about a specific particle) of a particle at any one point of time, Schrödinger provides you a method to depict the particle's wavefunction at any given point of time in the past or the future. So, if you know the quantum wavefunctions for each and every particle in the universe at present, you essentially have the potential to predict the future. Fundamentally, this statement carries much weight as it tells us that the concept of us having free will is just a misconception because the fate of the universe is entirely governed by the laws of physics. This explains the situation in the grandfather paradox where your inability to kill your grandfather can be attributed to the inevitable outcome, which has already been predicted by the laws of quantum physics regardless of you or anyone knowing the outcome or not.

Nevertheless, the instrumental question that fueled this entire debate in the first place, still remains – *is time travel to the past really possible?* The answer that most prominent physicists of today would give is a firm no, but in reality, no law of physics rules out the possibility of time travel to the past with utmost certainty. Moreover, some strong-willed physicists have proposed daunting ideas that stretch the limit of what is physically possible in the universe to make time travel in both directions equally permissible. Kurt Gödel, Einstein's colleague at Princeton, proposed the first mathematically possible solution for building a time machine, also known as Gödel's universe.

Gödel's analysis revolves around a fundamental idea of a rotating universe. In a previous analysis done by Scottish scientist W.J. van Stockum, Stockum showed that an infinitely long rotating cylinder along its infinitely long axis is capable of capturing space and time in a whirlpool and wrap them so much that if a rocket is guided in this whirlpool, the rocket will arrive at the starting point even before it embarked on its journey into the whirlpool. Long story short, the whirlpool's swirl around the cylinder will turn the direction of time on its head and open the possibility for objects to travel to the past. Gödel mathematically showed that similar to an infinitely long rotating cylinder, which we all agree is physically impossible to build, a rotating universe also allows for the possibility of time travel to the past. According to Gödel, if you follow precise mathematically defined trajectories in a universe that is rotating, spacetime would be wrapped so much that you would have the ability to arrive at the starting point of any journey before you even embark on it at any point of time. Roughly speaking, a rotating universe would itself act as a time machine, and, apart from being an expert cosmic geologist, if you know precisely how to drive your rocket in and out of the rocky terrain, you will master the properties of space and time.

But then, after further analyzing observations made regarding the possibility of a rotating universe, to Gödel's dismay had he known the results, physicists have now been established that our universe is not rotating after all. This limits the various dreamy possibilities provided by Gödel and his universe to only limited mathematical curiosities. Still, Gödel's solution was well received in the scientific community as it was the first of its kind proposal for a real mathematically possible time machine, and albeit it may not be practically viable, it has undoubtedly ignited excitement around the possibility of building a time machine capable of time-traveling in both the directions. Consequently, today, significant incremental work has been achieved in the domain of time travel, and one of the more plausible ideas which can be entertained for its likelihood comes from an extensive study of wormholes. An intriguing proposal shaped by Kip Thorne, a Nobel prize winning physicist who is a professor at the California Institute of Technology, and his students provide us with an elegant, simple and practically viable idea in contrast to the numerous other proposals that lay out a blueprint for building a potential cosmic time machine. This unprecedented proposal employs a stable and workable wormhole.

Wormhole Time Tunnels

A wormhole, as we have learned in the previous chapter *'holes and waves in the dark cosmos,'* is a cosmic bypass that truncates the amount of space across billions of light-years into only a couple of meters. The interesting ironical fact about these cosmic tunnels is that although they truncate space, wormholes do not occupy any space themselves like the conventional tunnel's, which we build here on Earth do. If say a wormhole exists between your city and your friend's city, which lies one hundred kilometers away, the wormhole will look something like how it has been depicted in figure 11.3. If you go right through this hole, you would be able to walk straight from your city to your friend's city. If instead, you decide to go around the hole with the expectation of finding your friend's city on the other side, you will never find his city behind the hole.

Moreover, for the same reason that wormholes do not occupy any space themselves, it is not possible to ever witness the tunnel-like structure depicted in figure 10.6 since it is only an artistic depiction and not an image of how the object may look like in reality. The truth is that a wormhole acts more of like a gate connecting two points in spacetime. This gate itself does not occupy any space, but if it is removed, the path which the gate offers by connecting the two points in spacetime will also vanish. This understanding of a wormhole not occupying space is crucial for us to understand Thorne's proposal of building a wormhole time machine.

Figure 11.3 An artistic impression shows how a wormhole might look like while connecting two different cities with one of its mouth placed in a city square.

Imagine that you are sitting in your home with one mouth of a wormhole in your room and the other mouth in your friend Sameer's room, who lives two streets across. You and Sameer are sitting in your respective rooms, chatting through the wormhole when Sameer decides to pay a visit to the Andromeda Galaxy to get one of the famous Andromedean space hoverboards for your birthday tomorrow. Sameer, as a good friend should, asks you to come along on the journey to visit Andromeda, but you, preparing for an important exam scheduled two days later, turn down the proposal. Instead, you persuade Sameer to load the mouth of the wormhole from his room into his spaceship so that when he reaches Andromeda galaxy, you may also have a look around by peeping through the wormhole. Sameer agrees to the condition, loads the wormhole into his spaceship, and leaves for Andromeda galaxy after arranging all space permits from the space police. At this point, you may think that as Sameer would travel across and in between different galaxies, the wormhole must stretch in length throughout the journey. This is the tricky part. Sameer does travel across space and his distance from you does keep on increasing, but alternatively, the wormhole is only a shortcut, and it is not this shortcut that travels through the conventional space that we all are familiar with. In this case, it is Sameer who travels through space, and the wormhole merely provides a gateway from your room to Sameer's spaceship wherever it may be across the cosmos. The explanation essentially means that the length of the wormhole remains *unchanged* throughout the trip. So, just like how you were chatting with Sameer before he embarked on his trip, you can still chat with him through the wormhole while he drives his spaceship in his transgalactic pursuit.

Furthermore, let us assume that Sameer travels at a speed of 99.9999999999999999 percent times the speed of light, and it takes him four hours to reach Andromeda Galaxy. When Sameer reaches Andromeda, he quickly buys the hoverboard and allows you to peep around the galaxy through the wormhole before heading back home. Four more hours later, after playing many rounds of tic-tac-toe and other games with you, Sameer reaches back home. While coming out of the cockpit of his spaceship, Sameer is shocked to find out that the date on the universal calendar he owns has suddenly jumped some six million years after the day he initially left for Andromeda galaxy. A second later, Sameer realizes what has happened. Because of the effects of time dilation, time slowed down for him in his spaceship, and the eight hours that elapsed for him were actually equivalent to some six million years back on Earth. Now, Sameer informs you that he is back home and has parked his spaceship on his lawn. Looking through the wormhole and Sameer's spaceship, you can clearly make out that Sameer is back on Earth, but looking out of your window, you cannot see Sameer's ship in his backyard which should clearly be visible as it always is. You ask Sameer, "I believe that you are back on Earth, but I am unable to spot your spaceship in your lawn, what is the deal?" Sameer then reminds you about the time dilation effect and says, "You are looking at the right spot in space but not at the right time. Due to special relativistic effects, I have not only traveled across space, but I have also traveled six million years into the future. Whereas, you simply stayed in the present by being at rest." To this you reply, "oh! I understand what has happened. Anyways, come through the wormhole and go back to your house or else you might get late for dinner." Sameer hops through the wormhole into your room, and scrambles back to his house.

Notice, it takes Sameer only a moment to travel through the wormhole, but he essentially travels six million years back in time. Therefore, the wormhole, in spirit, acts as a time machine, and anyone can confirm that. Your mom arrives in your room to call you for dinner and gets intrigued after seeing the wormhole in your room. She travels through the hole and talks to people on the street on the other side to realize that she has traveled six million years into the future. Your mom, scared by this time shift, scrambles back through the hole and ends up six million years back in the past. Essentially, the vital revelation that we learn from this experiment is that Sameer not only takes the wormhole through space, but he also takes it through time as well. In simple words, Sameer just *converted a space tunnel into a time tunnel; a wormhole has become a time hole.*

This process of building a time machine, however, has many limitations. The most important of the lot is that the time machine is not independent of the time difference in which it is locked. This means that the wormhole time machine, in this case, will only act as a time tunnel between two specific points in time, which will be six million years apart until their respective time frames are meddled with. Moreover, the wormhole time tunneling machine, that we have proposed, actually *does not allow for time-traveling prior to the time it is first conceived*. So, if the first wormhole time tunneling machine is built say 1000 years from now, many space tourists will flock to that moment to witness the conception of the very first time machine ever built, but any moment that has occurred before that specific moment will always remain inaccessible to these exceedingly eager space tourists.

Eventually, these limitations underscore the fundamental notion of time travel, which, according to our classical perception, must allow us to travel between any two points in time at will. Apart from this, building a wormhole is a headache on its own because unlike blowing up a mountain with dynamites to construct a tunnel, here we need to tear and join the fabric of spacetime, which cannot be achieved with a sharp kitchen knife and glue at our disposal. Besides, we have previously discussed that these cosmic tunnels require exotic matter to stabilize them, which are still to be fabricated. Everything put together, the invention of the very first two-way time tunnel is far from being realized today.

Theoretical Flights of Fantasy

Thinking about the possibility of time travel, Stephen Hawking and many others have frequently raised a pivotal question. Why, they ask, if a cosmic time machine has been built, have we not been visited by people claiming to be from the future? Surely, if not tourists, a historian at least must have got a grant to visit the past and document the construction of the first-ever atomic bomb, or the first-ever broadcast of reality television, or perhaps the first-ever voyage to outer space. Positively, there are reports of people self-identifying themselves to be time travelers and others claiming to have locked up people from the future in their lockers only to discover that they have vanished the very next day. But all in all, I am sure that as a neutral onlooker, you must not be judgmental by neither entertaining these claims nor profusely disapproving them. If in case you believe that we have not been visited by people from the future, perhaps you are implicitly implying the impossibility of building a time machine one day. But this paradoxical situation, under no circumstance, can prove that time travel to the past is impossible. We, as optimistic as we can be, must remain open-minded and believe in the possibility of past time travel until the laws of physics absolutely rule it out.

In a nutshell, a physicist's job can best be understood as an ironically tantalizing occupation that leads the employee towards a state of total confusion. Persevering through despondent days filled with destituteness, a theorist must embrace the realm of doubt while walking down the winding roads to clarity and coherence. To come up with ingenious, innovative, and inspired ideas that can potentially revolutionize the world of physics is a prize at the end of that long winding road, which is not kids' stuff that comes easy-breezy. Guided by hypotheses, calculations, approximations, failure, and to some extent, intuitiveness, the truth-seeking physicist has to overcome the resistance offered by nature to reveal its secrets. Idolizing Richard Feynman's famous words, *"I think nature's imagination is so much greater than man's, she's never going to let us relax,"* we do not know what future physical theories are hiding for us, what further fundamental rules we may learn of, or how our understanding of the cosmos might be superseded, once again.

Glossary

Absolute rest: a Newtonian concept; a vision of an unchanging, immovable, and omnipresent quantity independent of its contents.

accretion disc: a structure formed by diffused material in orbital motion around a star, typically black holes and neutron stars.

aether: a hypothetical substance that filled space to provide light a medium to propagate; discredited hypothesis.

Archimedes principle: the upward buoyant force that is exerted on a body immersed in a fluid is equal to the weight of the fluid that the body displaces.

baryonic matter: proposed dark matter composed of baryons; a hypothetical substance.

big bang: a cosmological model that explains the evolution of a hot and expanding universe from a moment after its birth.

big crunch: a hypothetical scenario in which the acceleration of the universe reverses and the universe collapses on itself to a cosmic scale of zero.

big rip: a hypothetical scenario in which the expansion of the universe which will progressively tear apart all of matter and spacetime at a certain point of time.

black dwarf: a theoretical stellar remnant of a white dwarf that does not emit heat or light.

black holes: a region of spacetime, which is so extremely dense that its gravitational acceleration does not allow even light particles to escape from it.

blue supergiants: scarce hot and bright stars that may or may not become red supergiants before exploding in a supernova.

classical physics: theories of physics such as Newton and Maxwell's theories which predate modern, more complete, and widely applicable theories.

cosmic microwave background radiation: remnant electromagnetic radiation from the early universe, which permeates all of space.

cosmological constant: the energy density of space, predicted by Einstein's equations, which drives the expansion of space.

cosmology: the study of the origin and evolution of the universe.

dark ages era: the era of the universe, dating probably between 380 million years to 1 billion years after the big bang, when luminous stars and galaxies were yet to form.

dark energy: a proposed form of energy which fills all of space; a more general impression of the cosmological constant.

dark matter: non-luminous matter which exerts gravitational influence while floating around in space.

dumb holes: a phenomenon in which phonons are unable to escape from a fluid that is moving faster than them in a local area, also called black hole analogs because of their similarity to black holes.

Einstein ring: deformation of the light from a source into a ring-like structure due to gravitational lensing effects.

electromagnetic force: one of the four fundamental forces which exerts its influence on particles with an electric charge.

equivalence principle: the equivalence of gravitational and inertial mass in general relativity.

event horizon: the luminous boundary of a black hole beyond which nothing, not even light particles, can escape.

exotic matter: hypothetical matter with exotic properties such as matter with negative density.

galaxy cluster: a structure that contains anywhere between hundreds to thousands of galaxies that are bound together by the gravitational force of dark matter.

gamma ray burst: brightest most powerful explosions from astrophysical sources that emit gamma radiations.

general relativity: Einstein's theory of gravity, which invokes the concept of curvature of space and time.

geodesic path: the shortest distance between two points in Riemannian manifold (four-dimensional spacetime).

gravitational lensing: the phenomenon of bending of light around the distribution of matter between the light source and the observer.

gravitational redshift: a phenomenon in which time runs slower in the presence of a gravitational field as observed by an external observer.

inflationary cosmology: a cosmological theory that explains a brief but enormous expansion of space in the very early universe.

inflationary era: the first era of the universe predominated by the inflationary field, which drove the initial spatial expansion.

inner planets: the group of Mercury, Venus, Earth, and Mars is called the inner planets.

length contraction: the phenomenon in which a moving length is measured to be shorter than its proper length.

light clock: a clock designed to work by flashing a light ray between two parallel mirrors while measuring the number of flashes occurred in a period of time.

LIGO: a gravitation wave detector in the US used as an observatory to develop gravitational waves as an astronomical tool to study space.

magnetars: a type of neutron star with powerful magnetic fields that power high-energy gamma rays.

mid-size stars: stars which usually have a life cycle of about ten billion years and whose masses lie in the range of 0.4 to 8 solar masses.

muons: an elementary particle similar to an electron.

nebula: a giant cloud of dust and gas in space, which is usually a remnant of a supernova explosion.

neutron stars: a stellar object mainly composed of densely packed neutrons with a tiny radius of typically 20 miles.

observable universe: a spherical region of the universe comprising of all matter than can be physically observed from Earth.

particle accelerator: a research tool used to study particle physics by smashing elementary particles together at high speeds.

phonon: a sound particle which carries the information of a sound.

photoelectric effect: the phenomenon of emission of electrons when high-speed photons strike a material.

photon: an elementary light particle.

planetary nebula: a glowing shell of expanding gas emitted by red giants late in their lives.

plasma era: the era of the universe during which matter existed in the fourth state of matter, plasma.

protostar: a very young star still gathering mass from a molecular cloud.

pulsars: a highly magnetized rotating neutron star that seems to periodically emit electromagnetic radiation.

quantum mechanics: a theory of physics that explains the nature and working of subatomic particles at micro scales.

quantum wave function: it is a mathematical description of the state of an isolated subatomic particle, which is given by quantum mechanics.

quarks: an elementary particle that is the fundamental constituent of all visible matter.

quasar: an extremely luminous active galactic nuclei of a supermassive black hole that emits powerful electromagnetic radiations.

red dwarfs: small stars whose mass is usually lower than 0.4 solar masses.

red giant: a luminous mid-sized star in the last phase of its fuel-burning cycle.

red supergiants: giant stars that have a typical life cycle of around a million years and are usually more massive than 8 solar masses.

relativistic mass: a relativistic change in the mass of a body as it approaches near light speeds.

relativity of simultaneity: the concept that the possibility of two spatially separate events occurring simultaneously depends upon the observer's frame of reference.

spacetime: the property that time and three-dimensional space are fused together to form a four-dimensional world.

special relativity: Einstein's physical theory which establishes a relationship between space and time.

stellar era: the fifth era of the universe, which was predominantly driven by the production and evolution of stars, galaxies, and other astrophysical objects.

strong force: a fundamental force of nature that binds quarks and nucleons (protons and neutrons) together.

supernova: a massive, luminous stellar explosion that marks the last evolutionary stage of a massive star.

tachyons: hypothetical particles that are believed to travel at speeds higher than the speed of light in an alternate reality.

time dilation: the phenomenon of special relativity in which time runs slower for moving objects.

vacuum: a region of space devoid of matter.

white dwarf: a dense star which is a stellar remnant of a mid-sized star after which has exhausted its fuel.

white hole: a hypothetical astrophysical object which is the complete reverse of a black hole; it comprises a region of spacetime that emits matter and probably time.

wormhole: a hypothetical astrophysical object which speculatively connects two spatially separate points in spacetime through an alternate shorter path.

References

- Wikipedia
- The Fabric of Spacetime by Brian Greene
- Relativity by Albert Einstein
- A Brief History of Time by Stephen Hawking
- Ideas and Opinions by Albert Einstein
- physics.stackexchange.com
- www.goodreads.com
- www.brainyquote.com

www.ingramcontent.com/pod-product-compliance
Lightning Source LLC
Chambersburg PA
CBHW030919180526
45163CB00002B/397